FUSING
Technology

GERRY COOKLIN
*SHENKAR COLLEGE OF
TEXTILE TECHNOLOGY AND FASHION*

*THE TEXTILE INSTITUTE
10 Blackfriars Street
Manchester M3 5DR
United Kingdom*

THE TEXTILE INSTITUTE
10 Blackfriars Street
Manchester M3 5DR
United Kingdom

First Published 1990

ISBN 1 870812 22 0 Paperback
1 870812 20 4 Hardback

Printed by Trafford Press, 418 Chester Road, Manchester M16 9HP

FUSING TECHNOLOGY

PREFACE

A garment with immediate sales appeal is an essential objective for every clothing manufacturer. However, one of the most important components of nearly every garment is that which has little or no sales appeal because it is designed not to be seen: the fusible interlining.

Fusibles were first introduced on a large scale in the 1950s, and they started what has since been called the third clothing revolution, the other two being the discovery of weaving and the invention of the sewing machine. Since then, the fusing of garment components has become an integral part of clothing technology. Today, it is widely accepted that garment quality and durability are inexorably bound to the quality and application of the fusibles used.

This handbook was originally compiled for students of clothing technology, and the objective was to present a restated selection of the literature available under one cover.

I should like to extend sincere thanks to the late Professor Gad Alon and my colleague Engineer Sam Shaltiel for their help and encouragement during the preparation of this handbook. I also wish to express my appreciation to Dr C M C Dorkin for her comments on the manuscript.

Gerry Cooklin
Shenkar College of Textile Technology and Fashion

CONTENTS

		Page
1.	Introduction	7
2.	Base Cloths (Substrates)	14
3.	Resins	23
4.	Coating Systems	33
5.	Selection of Fusibles	37
6.	The Methods and Components of Fusing	41
7.	Machinery and Equipment	50
8.	Working Methods	74
9.	The Control of Quality	81
10.	Glossary of Terms	88
11.	Bibliography and Acknowledgements	92

Graphics by Etti Tovianah

1. INTRODUCTION

The first known use of interlining for stiffening clothes was by the ancient Egyptians in about 3000 - 4000BC. The nobles of that time wished to present an image of power and authority, and, among other things, this was done by wearing enlarged and exaggerated clothes and hats (*Fig 1*). Large parts of their garments were stiffened with heavy fabrics interwoven with metallic threads or strips, frequently made from gold or other precious metals. At different periods of history, the elaboration of garment styles called for stiffening materials, and, for many centuries, buckram and linen interlinings were used for this purpose.

Until the nineteenth century, buckram, a coarse cotton fabric stiffened with glue, was widely used as an interlining. This material was gradually replaced by interlinings that included animal hair in their construction. The most widely used hair cloth was that made from the hair of horse tails, and, with various refinements, this was used as an interlining for more than 150 years.

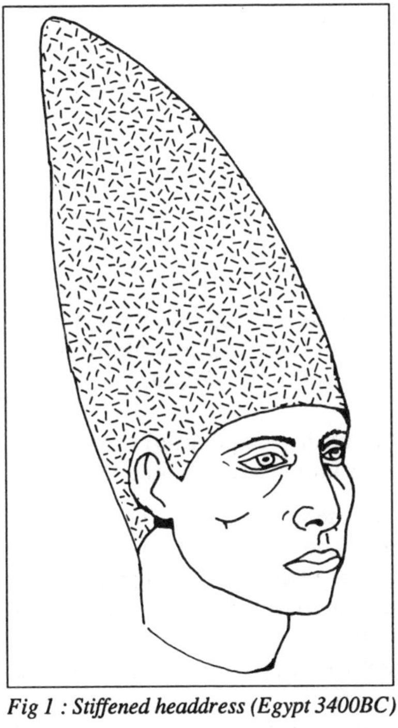

Fig 1 : Stiffened headdress (Egypt 3400BC)

The process of attaching an interlining, especially to the forepart of a jacket or coat, required a high degree of tailoring skill in order to ensure that the top-cloth remained free of distortion. A tailor performing this operation would sew rows of tacking (basting) stitches to hold the interlining and cloth together and then finely hand-sew the lapels with countless rows of small stitches, called padding. Other parts of the interlining were fastened to seams, darts, and pockets by carefully sewn tacks. Finally, the forepart with the attached interlining was hand-pressed with great care to ensure that the two layers were perfectly matched (*Fig 2*).

Where interlinings had to be attached to small compo-

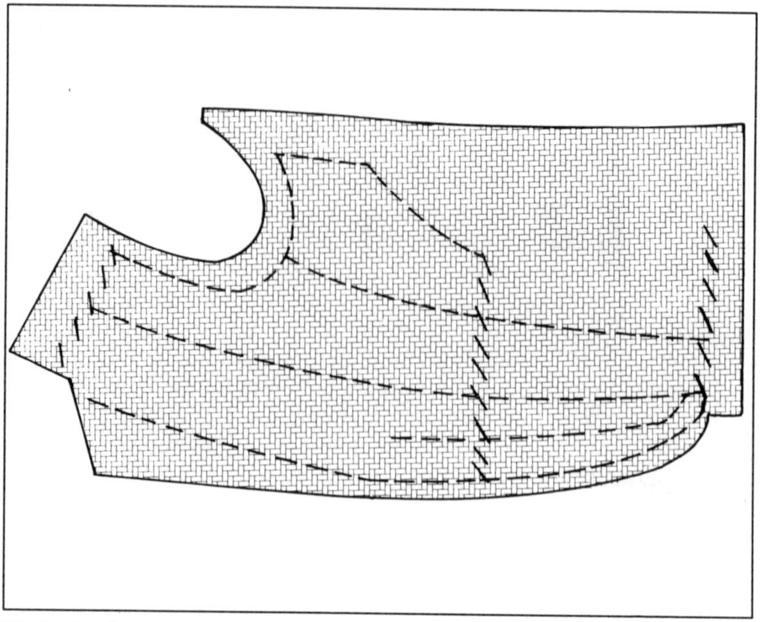

Fig 2 : Interlining attached by hand (man's jacket forepart)

nents, the usual method was to sew the two plies together by a runstitch on the edge of the components.

Towards the end of the 19th century, the use of sewing machines became more widespread, and many hand-sewing operations were gradually replaced by regular or special types of sewing machines. The machines that were developed for attaching the top-cloth to the interlining were as follows.

(1) The Plonker (Joker) This machine duplicated the hand-sewing operation whereby the top-cloth and interlining were held together prior to the final attachment (*Fig 3*).

(2) The Padding Machine This machine duplicated the rows of stitches that were hand-sewn on lapels and collars (*Fig 4*).

Although the use of machines for these processes considerably reduced the time element involved, the quality of these critical operations was still dependent, to a large degree, on highly skilled labour. Consequently, quality was inconsistent, as is usual with hand- or operator-controlled sewing.

This situation underwent a revolutionary change just after the end of the Second World War in 1945.

At this time, because of the enormous demand for all types of

Fig 3 : Plonking machine (joker)

Fig 4 : Padding machine

clothing and the soaring costs of natural fibres, textile and clothing manufacturers were forced to find new and cheaper cloths. To this end, they were helped by the discovery in 1938 of nylon, which was the first synthetic fibre successfully produced in a laboratory. Nylon, together with polyester fibre (discovered a few years later), enabled the textile industry to produce fabrics having the characteristics necessary for clothing.

The introduction of these new fabrics brought with it a host of problems, and new technologies had to be developed for the cutting, sewing, and pressing of garments made from synthetic-fibre-based cloths. These problems were gradually overcome, but there still remained the most problematic area of all, interlinings.

Before the large-scale use of synthetic fibres, most of the cloths used for outerwear were of wool, and this allowed for a certain degree of manipulation during the attaching and pressing of interlinings. However, these new cloths were of a more critical nature, and consequently the interlinings used had to be perfectly matched to the characteristics of the top-cloth and, because of labour costs, be capable of being attached by a completely machine-controlled process. These conditions made it essential to develop different types of interlinings and the methods of attaching them to the top-cloth.

The solution was fusible interlinings.

The term 'fusible interlining' is generally defined as a base fabric having a deposit of thermoplastic adhesive resin (usually on one surface only), which can be bonded to another fabric by the application of heat and pressure. The development of fusibles was a fairly rapid process and required the combination of three separate technologies:

(1) textile: to produce the necessary types of base cloths;

(2) resin: to produce thermoplastic adhesives for heat bonding;

(3) engineering: to produce machines that would fuse the interlining to the top-cloth.

These new fusibles were very different from the experimental materials in existence up to this time. For some years prior to 1950, fusibles had been used on a limited scale for stiffening the collars and cuffs of men's shirts. These applications were not very successful because of the tendency of the fusibles to separate (delaminate) from the top-cloth after washing or dry-cleaning. Early in 1951, Dr. Harold Rose developed the first practical fusibles for clothing and thus started the third revolution in the history of clothing.

Today, the use of fusibles for mass-produced clothing instead

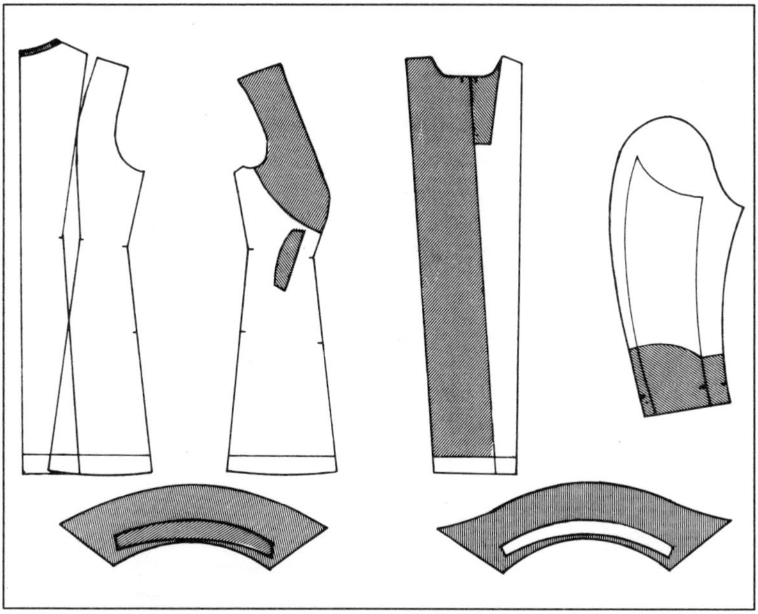

Fig 5 : Fusible application: lady's coat

Fig 6 : Fusible applicaton: man's jacket

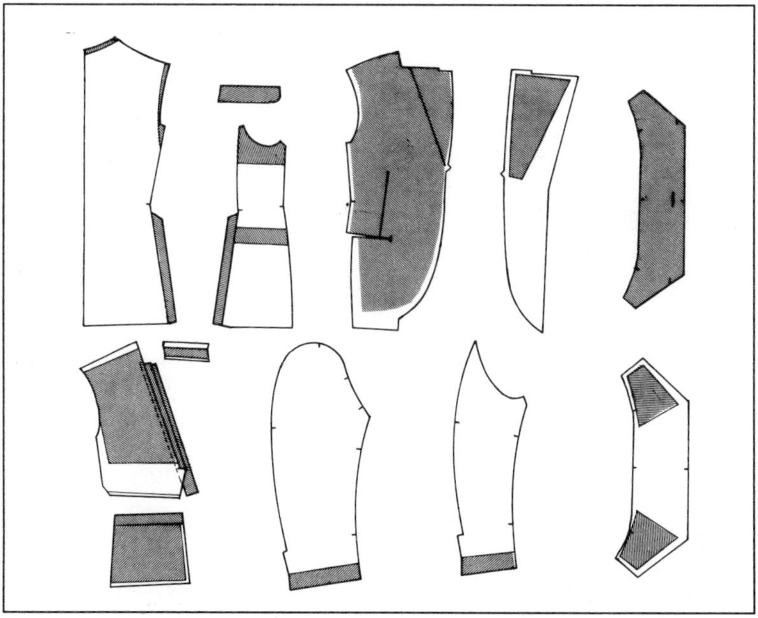

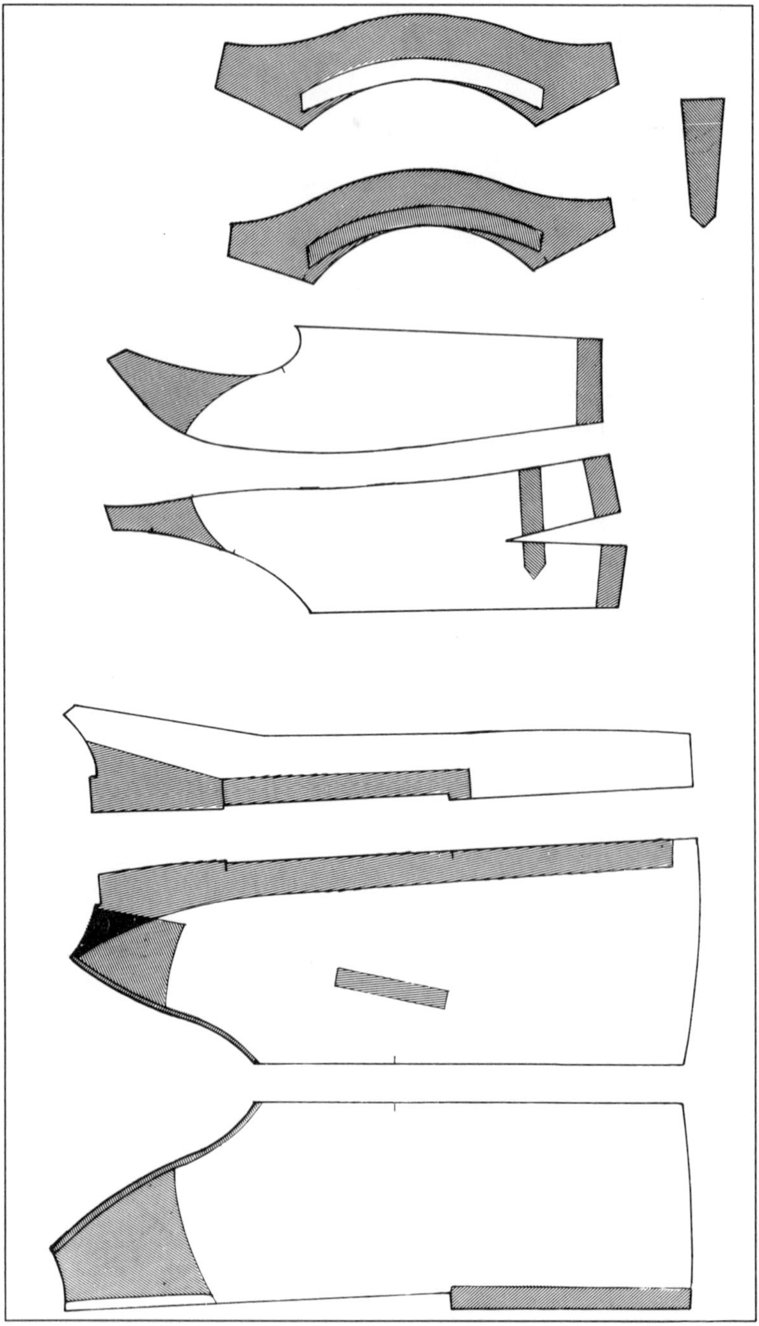

Fig 7 : Fusible application : rainwear

of the traditional sew-in type of interlining offers many advantages to both the manufacturer and the wearer. These superiorities can be summed up as follows.

For the manufacturer

(a) Garments can be produced with a consistent quality of interlining construction.

(b) Time and labour costs are saved by eliminating the hand and machine operations required to attach the interlining to the top-cloth.

(c) Handling during production, especially of small components, is facilitated.

(d) Differential shrinkage between the top-cloth and interlining is reduced to controllable proportions.

(e) Stitch pucker and the consequent distortion of sewn parts are reduced.

(f) The garment has a cleaner and fresher appearance upon its initial presentation.

For the wearer

(a) Garment durability is increased.

(b) The handle of the cloth is only very slightly modified by the interlining.

(c) The garment retains its original shape after repeated dry-cleaning or washing.

(d) Crease-recovery time is considerably reduced.

Thus, from all points of view, the use of fusibles for mass-produced clothes offers many advantages in terms of garment production and garment performance. Typical examples of the application of fusibles to various types of garments are shown in *Figures 5, 6,* and *7.*

2. BASE CLOTHS (SUBSTRATES)

The base cloth, also called the substrate, is an interlining material onto which the thermoplastic resin is coated, sprayed, or printed. Base cloths can be produced in a variety of woven, knitted, and nonwoven forms, each type having a specific application, according to the planned end-use. It is difficult to do more than generalize on practical comparisons between the different types of base cloths, since their properties overlap to a considerable extent. The base cloth is only one component of the fusible, and therefore can only be evaluated in relation to the other parameters concerned. *(This subject is examined in more detail in Chapter 5*

Irrespective of the type used, the base cloth influences the following characteristics in the finished garment:

(a) handle and bulk;

(b) shape retention;

(c) shrinkage control;

(d) crease-recovery;

(e) appearance in wear;

(f) appearance after dry-cleaning or washing;

(g) durability

In addition, the final cost of the garment is also influenced by the type and amount of fusibles used in its construction.

The general properties and characteristics of each major group of substrates are as described below.

(A) Woven Interlinings

Weaving is the process of forming a fabric on a loom by interlacing the warp and weft yarns with one another. Woven substrates can be produced from animal hair, viscose, cotton, polyester fibre, acrylic fibre, wool, or in any number of combinations of these fibres.

Because of their construction, woven fabrics are not easily distorted by wear or cleaning, and, as a result, they exert maximum control on shrinkage and shape retention. However, these properties are not conducive to a soft and natural handle in the finished garment. Special weaves, such as twills or fabrics with fine warps and coarse wefts, are recent developments that somewhat improve the handle and bulk.

The possibility of using different types of yarns for the warp and weft enables substrates to be produced with special properties. An outstanding example of this application is an interlining with a soft handle in the warp and a firm handle in the weft. This interlining is specifically produced for the fronts of men's jackets and provides a firm basis for the length of the front and also prevents the lateral sections across the chest and shoulder from collapsing. Another recent development, utilizing the same principle, is the graduated or polyzonal interlining, which is woven with three different zones along the warp. Each zone possesses different properties in terms of handle, drape, and resilience. A front fusible cut from a polyzonal base cloth (*Fig 8*) will possess all the necessary attributes for a jacket front:

(a) a highly resilient shoulder area at the top;

(b) a resilient, but softer, chest area in the centre;

(c) a super-soft, yet crease-resistant, lower forepart at the bottom.

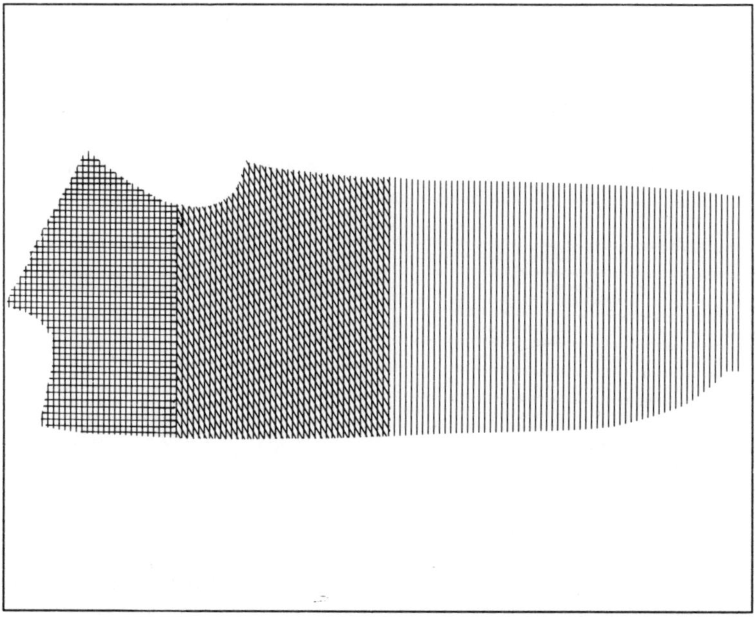

Fig 8 : Polyzonal fusible

This type of construction makes the garment lighter, improves its appearance, and significantly reduces labour costs in comparison with the traditional method of chest-area construction. Woven interlinings can be used for all types of garments where strength, stability, and good draping qualities are required. Their major disadvantage is the relatively high cost in comparison with the other types available.

(B) Knitted Interlinings

The growing use of knitted fabrics for men's and women's outerwear led to the introduction of knitted interlinings in the early 1960s. These interlinings, because of their construction, provided a considerable degree of elasticity to the fused components by yielding easily to body and limb movements.

The first products were warp-knitted (*Fig 9*), and it was not until the introduction of weft-inserted fibres into the warp-knitted construction that this type of interlining became widely accepted as a fusible for woven cloths (*Fig 10*). These substrates are basically a warp-knitted fabric with an inlaid weft yarn, usually of viscose, and are raised to give a soft, lofty handle. The advantage of a weft-inserted interlining is that it has a natural handle while providing resilience in the weft direction, i.e., around the body. The newest weft-inserted types include a stretch property in the weft fibre that provides extensibility in this direction and thus makes it more compatible with the outer fabric. A further development has been a polyzonal type graduated to provide the necessary firmness on each section of the front.

The warp-knitted interlinings are widely used for women's light clothing, such as blouses and dresses made of crêpe-de-

Fig 9 : Warp-knitted base cloth

Fig 10 : Weft-knitted base cloth

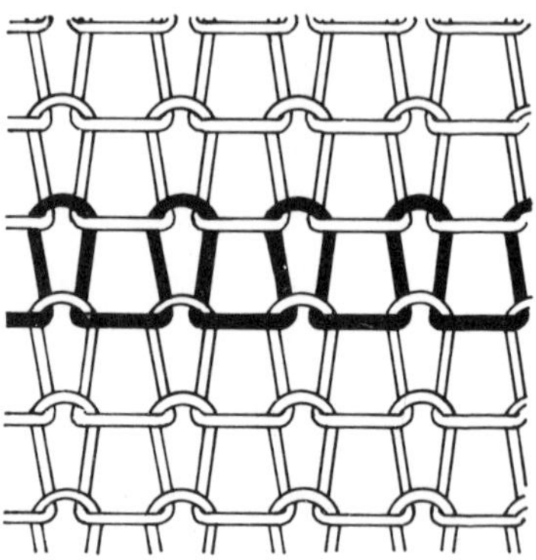

chine, georgette, and polyester-fibre yarns. These types of fusible are available in all weights and colours.

The fibres used for knitted substrates are usually man-made and include polyester fibre, polyamide fibre, acrylic fibre, and viscose or any combination of these. Since the knitting process is generally faster than weaving, the cost of knitted substrates is somewhat lower than that of an equivalent woven product.

(C) Nonwoven Interlinings

The nonwoven interlinings used by clothing manufacturers originate from the paper industry, and their development was brought about by shortages of traditional fabrics. Initially, these fabrics were used for industrial purposes, such as filter cloths, but improvements in chemical adhesives and the associated machinery enabled manufacturers to produce nonwoven fabrics that were acceptable to clothing producers.

A nonwoven substrate is a series or mixture of fibres, held together at bond sites, and its end-performance is determined by the following factors:

(1) fibres: whether natural, synthetic, or a mixture of both;

(2) web formation: how the fibres are orientated in the substrate;

(3) bonding: the process whereby the fibres are locked together.

(1) Fibres

Owing to the cost of natural fibres, the great majority of nonwoven substrates are made from man-made fibres, as follows.

(i) Viscose: This fibre is made from a cellulose base and is sometimes referred to as regenerated-cellulose fibre.

(ii) Nylon: Nylon is a generic term and refers to a class of fibres that can be modified to meet particular requirements.

(iii) Polyester fibre: During the research that led to the discovery of the polyamides used for the manufacture of nylon, another related type of fibre-forming polymer emerged, which is known as polyester.

(iv) Acrylic fibre: This fibre is far more amenable to ordinary dyeing processes than other synthetic fibres of a similar type.

(2) Web Formation

The basic methods of orientating the fibres that form the substrate or web are (*Fig 11*):

(a) dry-laid web formation;

(b) wet-laid web formation;

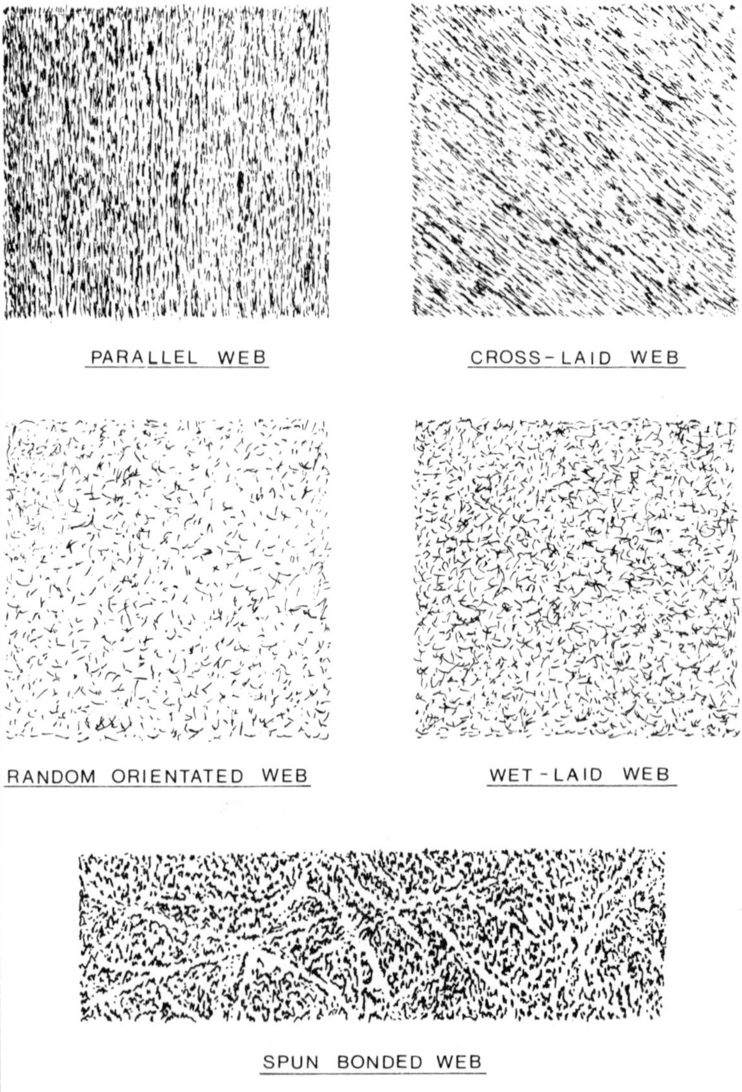

Fig 11 : Nonwoven web types

(c) spunbonding.

(a) Dry-laid Webs The fibre web can be formed mechanically or by air by using the following methods for the formation and orientation of the fibres.

(i) Mechanical Method This method passes the fibres through a card (an opening and orientating machine) and deposits them onto a moving conveyor. The angle of the card in relation to the direction in which the conveyor moves enables the following types of webs to be produced.

Parallel The card is parallel with the direction in which the conveyor moves.

Transverse This web is formed by placing the card at right angles to the direction of the conveyor.

Cross-laid A combination of the parallel and transverse methods of web formation.

(ii) Air-laid Method An air-laid web is formed by blowing single fibres onto a screen so that the fibres are orientated in random positions. The resultant web has neither parallel nor transverse orientation.

(b) Wet-laid Webs The wet-laid process is basically that used in the manufacture of paper. The fibres are suspended in liquid and are fed onto a conveyor screen, thereby forming the web. Suction devices under the screen remove the liquid, which results in the fibres orientating in a three-dimensional formation.

(c) Spunbonded Webs A spunbonded web is formed by melting and extruding the polymer through a system of rotating spinnerets. The extruded filaments are drawn by air onto a conveyor to form the web.

The dry-laid web is the most generally used because it achieves a good balance between handle, strength, and elasticity. Wet-laid formations tend to be paper-like and 'boardy' but have some applications where high strength is required. The spunbonded webs, although strong and elastic, are used mainly for sew-in interlinings.

(3) Bonding

As a result of the web-forming process, the fibres are orientated but not locked together. Bonding is the process that locks the fibres together.

(1) Mechanical Bonding The most widely used method of mechanical bonding is the needle-punching or felting technique, and the substrates produced by this technique are referred to as needle-punched. The process consists in piercing the web with barbed needles to lock the fibres together and thus form a stronger fabric. An increase in needling cycles will generally increase the strength of the fabric. Another

mechanical-bonding process locks the fibres together by spraying the web with high-pressure water jets while it is moving on a perforated conveyor. The finished fabric can be given a woven or knitted appearance by the different patterns of perforations on the conveyor.

Chemical Bonding The chemical adhesives widely used for fibre-bonding are as follows.

Nitrile Rubber This base gives a substrate with good draping qualities; however, the interlining tends to 'yellow' with time.

Acrylic Although this produces a pure white substrate, the resultant interlining tends to be somewhat stiff.

Styrene Butadiene Styrene-based products are generally used as binders for low-cost nonwoven fabrics, and the resultant fusibles are mainly limited to wash-and-wear products.

It is important that the chemical-bonding agent not only preserves the formation of the fibres for the life of the garment but also imparts softness and drape to the fabric. The methods of chemical bonding are as follows.

(i) *Impregnation* The web is saturated by being passed through a bath of the binder, squeezed between two rollers, and then dried at the temperature appropriate for bonding the fibres.

(ii) *Spraying* The web is transported on a conveyor and sprayed with the binder. Correct penetration is achieved by the use of a powerful vacuum positioned under the conveyor.

(iii) *Foaming* The web is saturated with a foamed binder that allows the fibre to interlock in a consistent form. This gives the substrate superior elasticity and crease-recovery characteristics.

(iv) *Print Bonding* An engraved roller applies the binder as the web passes under the saturated roller. The fibres are interlocked by the pattern of the roller; this method is used mainly for parallel-type webs.

(v) *Thermal Bonding* This is achieved by blending a low-melting-point adhesive powder with the web fibres. When the web is heated, the thermoplastic adhesive bonds the fibre structure together.

(vi) *Heat Welding* This method utilizes the thermoplastic properties of the fibres themselves. The fibres are Hetero-filaments of the core-sheath type (*Fig 12*) in which the core has a high melting point and the sheath has a low melting point. When a web of these fibres is

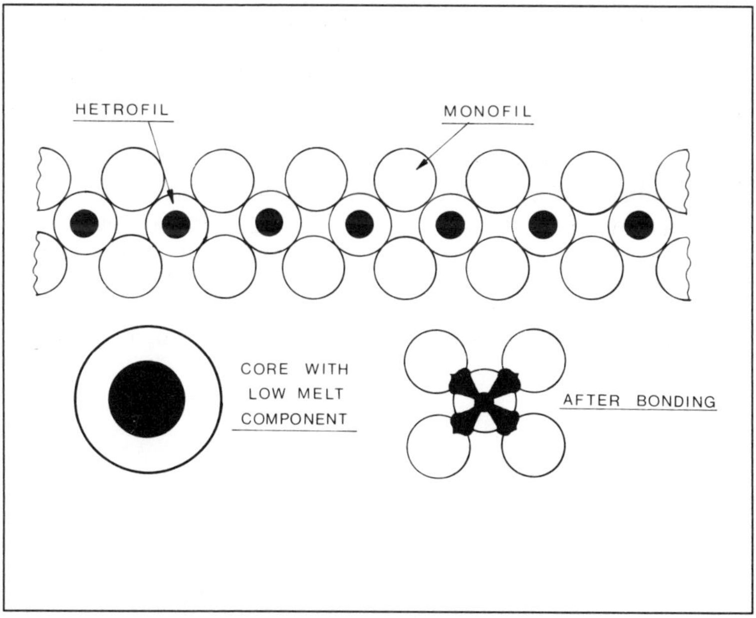

Fig 12 : Binding with Heterofil fibres

heated to the melting point of the sheath, bonding is effected at the site where the fibres are in contact. This type of bonding gives a softer handle than that achieved by chemical binding agents owing to the ratio between the Heterofil and monofil fibres. A typical monofil fibre construction that has been chemically bonded contains a larger proportion of binder material, which tends to 'harden' the handle of the interlining (*Fig 13*).

Fig 13 : Binding with monofil fibres

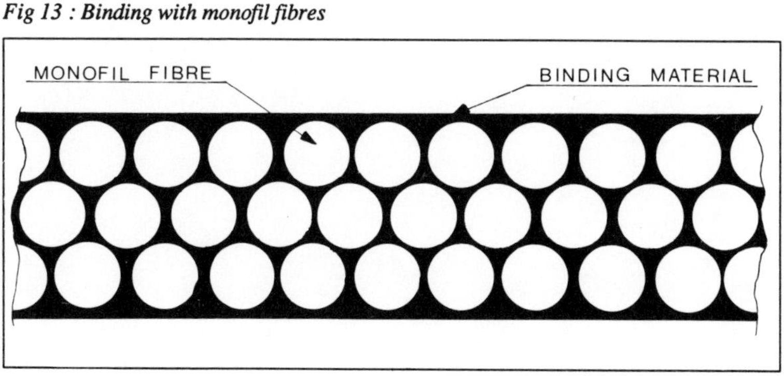

(vii)Ultrasonic The web is transported under an ultrasonic transducer, and a pattern of ultrasonic beams melts the fibres and bonds the structure. This newly developed method creates fabric patterns similar to those obtained by weaving.

Nonwoven structures are finished in much the same way as woven fabrics. These processes include calendering, dyeing, and printing. Nonwoven substrates, because of their method of fabrication, are generally less expensive than woven or knitted fabrics.

Each of the three basic substrates described in this chapter has its own advantages and disadvantages, and the factors governing the choice of appropriate substrates will be examined in*Chapter 5.*

3. RESINS

These are the materials applied to the base cloth, and, when subjected to heat and pressure, they become the sole bonding agent between the top-cloth and the interlining. Whereas these adhesives have customarily been described as fusible resins, this term is something of a misnomer. The word 'resin' usually refers to a natural product formed by secretion in almost all trees and plants. Today, no naturally occurring resins are used for interlinings, but an extensive and varied range of thermoplastic synthetic resins are employed.

Thermoplasticity, or change with heat, is the basis of all fusible interlinings. The adhesive coating in its cold state is not adhesive but, when subjected to heat and pressure, melts and creates a bond between the substrate and top-cloth. Upon cooling, the adhesive reverts to its original solid state, and, by virtue of penetration, it maintains the bond between the two fabrics (*Fig 14*).

Whereas there are numerous thermoplasatic resins, their suitability for interlinings is restricted by the following conditions.

(a) Upper-limit Fusing Temperature This must be below the temperature that would damage the top-cloth or cause changes in colour. Theoretically, this temperature would vary according to the composition of the top-cloth. However, in practice, it is rare for this temperature to exceed 174°C (345°F).

(b) Lower-limit Fusing Temperature This temperature is limited because of the necessity to ensure the adequacy of the bonding required in order to withstand the handling to which fused components are subjected during production. For most cloths, this lower limit is about 110°C (230°F), although leather and suede garments would require much lower temperatures.

(c) Cleanability The adhesive qualities of the resin have to withstand washing and/or dry-cleaning throughout the life of the garment.

(d) Thermoplasticity This particular property of the resin must be such that the change of heat combined with the correct pressure is sufficient for the resin to penetrate the top-cloth and form an efficient bond.

(e) Colour For the majority of end-uses, the resin must be white or transparent unless there are specific colour requirements.

(f) Handle The resin must contribute to the desired handle of the final laminate between the top-cloth and interlining.

(g) Safety For obvious reasons, the resin must be harmless in processing and end-use.

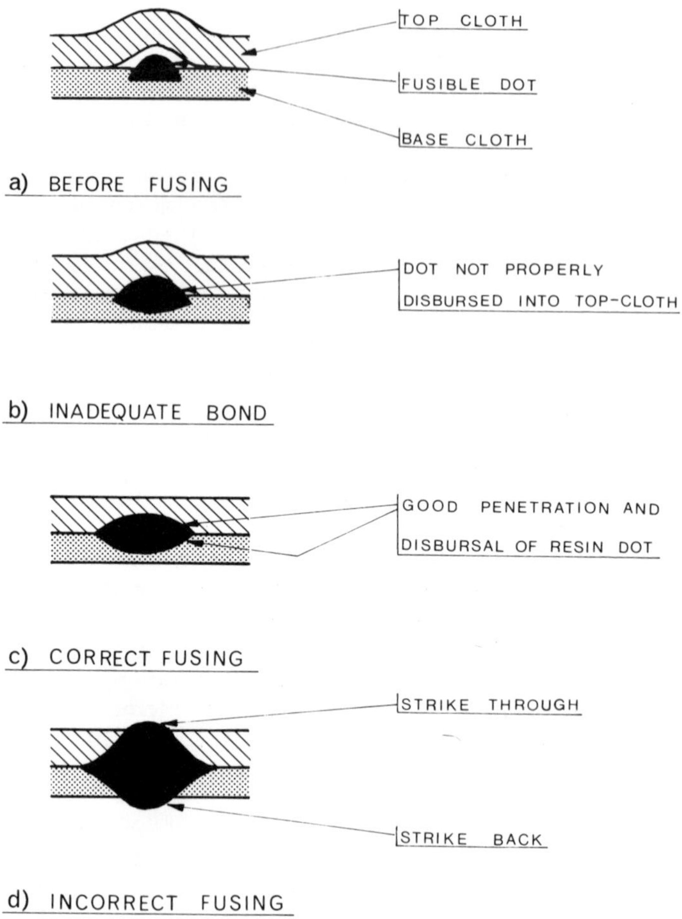

Fig 14 : Disbursal and penetration of resin

Resins are applied to substrates in three different densities; low, medium, and high, where the degree of density refers to the actual mass per unit volume of the resin material itself. It is this physical density to which the melting point of the resin and its resistance to dry-cleaning solvents are related. As a general rule, the higher the density, the greater is the resistance to dry-cleaning solvents. The main characteristics of the thermosplastic resins most widely used for fusible interlining are as follows.

(i) Polyethylene Ethylene, when polymerized under high pressures and temperatures with the aid of suitable catalysts, produces a polymer known as polyethylene. This can be converted into synthetic fibres as well as thermoplastic resins. Polyethylene-coated interlinings with low and medium densities are used for wash-and-wear garments, and the high-density range is both washable and dry-cleanable.

(ii) Polyamides: Polyamide resins, which were developed in the mid-1970s, are generally terpolymers made by reacting the basic ingredients of nylon 6, 6.6, and 11 or 12. By varying the constituents, a wide range of such properties as melting range and flexibility can be achieved.

All polyamide resins produce full, dry-cleanable bonds, but the lower the recommended fusing temperature, the lower is the resistance to hot washings. With the exception of some special resins, two classes of polyamides are generally used for fusible interlinings:

(1) Higher-melting range: Dry-cleanable and washable up to 60°C.

(2) Lower-melting range: Dry-cleanable only.

These melting ranges reflect the different temperatures achieved in the normal washing and dry-cleaning processes.

(iii) Polyester This is a relatively new development and is similar in performance to polyamide resins. It has a wide range of fusing temperatures (120---65°C) and is washable and dry-cleanable. Polyester resins are generally cheaper than polyamides and are now becoming widely used for coating the interlinings of men's jackets and coats.

(iv) PVC (Poly(vinyl chloride)) This resin is used extensively for siliconized rainwear fabrics. PVC requires softening with additives called plasticizers that are soluble in dry-cleaning solvents; consequently, the fused components have a tendency to stiffen when dry-cleaned. Fusing temperatures of this resin can be varied by adjusting the amount or type of plasticizer used in its formation.

(v) *Resin Blends*: Mixtures of the above-mentioned products are sometimes used with or without the addition of other resins to produce specific results, such as improved washability, better low-temperature fusing, etc. Three of the more widely used resin blends are as follows.

(a) *Plasticized Cellulose Acetate* This was the first multipurpose fusing resin and is still widely used for a great variety of fusible interlinings. It is both washable and dry-cleanable in the fusing-temperature range of 120-165°C.

(b) *Phenolic Resins* These are cheap resins with poor dry-cleanability and have the disadvantage of darkening in colour during processing and/or exposure to strong light. This prevents their use on lightweight cloths.

(c) *Plasticized Poly(vinyl acetate)* This resin is mainly used for leather and fur garments because it bonds at very low temperatures and pressures. It is not dry-cleanable and has very limited washability.

Some other resin blends are:

(i) *partially hydrolysed poly(vinyl acetate):* dry-cleanable, low washing-resistance;

(ii) *partially hydrolysed ethylene poly(vinyl acetate):* dry-cleanable and washable;

(iii) *ethylene vinyl acetate:* washable only;

(iv) *ethylene vinyl acetate terpolymers:* washable only.

To decrease the over-all costs of resin coatings, dry-cleanable resin powders are often blended with low-density polyethylene powder. This mixture provides a coating that has an acceptable initial bond that will have good washability combined with a reasonable measure of resistance to dry-cleaning.

Irrespective of the substrate and resin being used, all fusible interlinings require temperature, pressure, and time to ensure their correct application to the top-cloth. There must be a source of heat that will change the thermoplastic resins from a dry, non-tacky solid to a tacky, viscous adhesive. This heat source must pass through the substrate layer to the resin itself and then on to the top cloth.

With the application of pressure for a predetermined period of time, bonding will occur. On cooling, the bond is retained, and the resin reverts to its original solid state.

Every fusible coating has an optimum fusing temperature at which it will give the best results. If the temperature achieved is not sufficient to effect the correct change in the resin, then extra time or higher pressures will not compensate for lower temperatures. On the other hand, temperatures that are too high or too long a time can also cause problems. If the resin becomes too fluid, then when under pressure it will strike-back or strike-through the top-cloth or substrate or both.

To sum up, each resin type has its own specific characteristics, and these have to be considered in relation to:

(a) the style of garment being produced;
(b) the top-cloth and the handle required;
(c) the fusing equipment available in the factory;
(d) the base cloth of the fusible;
(e) the over-all cost;
(f) the durability to washing or dry-cleaning or both.

The main criterion for selecting a particular resin is that it should achieve the best results from the viewpoints of the designer, the technologist, and the end-user. This is not the same as achieving the best result at the lowest price. Consequently, comments based on resin types only are of limited value. However, there are some basic properties and characteristics of resin types that influence choice, and these are summarized in the following tables:

Table 1: General Properties of Resins;
Table 2: Children's-wear Applications;
Table 3: Men's-wear Applications;
Table 4: Ladies'-wear Applications;
Table 5: Other Applications (Rainwear, Leather, and Fur).

Table 1: General Properties of Resins

Resin Type	Fusing System			Durability		Over-all Cost
	Steam Press	Electric Flat-bed	Continuous-fusing	Washability	Dry-cleanability	
Polyethylene *(low-density)*	Limitations	Very suitable	Limitations	Suitable	Limitations	Low
Polyethylene *(high-density)*	Not recommended	Very suitable	Suitable	Very suitable	Very suitable	Medium
Polyamide copolymer	Suitable	Very suitable	Very suitable	Limitations	Very suitable	High
Polyamide terpolymer	Suitable	Very suitable	Very suitable	Limitations	Very suitable	High
Poly(vinyl chloride)	Not recommened	Very suitable	Very suitable	Suitable	Very suitable	Medium
Plasticized cellulose chloride	Limitations	Very suitable	Very suitable	Suitable	Suitable	Medium
Phenolic resins	Limitations	Very suitable	Very suitable	Limitations	Suitable	Medium
Plasticized poly(vinyl acetate)	Suitable	Very suitable	Very suitable	Not recommened	Limitations	Medium

Table 2 :Children's-wear Applications

Article	Resin / Area	Low-density Polyethylene A	Medium-/High-density Polyethylene B	Polyamides C	PVC D	Cellulose Acetate E	Phenolics F
Jackets	Chest-piece	-	Suitable	Suitable	Suitable	Suitable	Suitable
	Front	-	Suitable	Suitable	Suitable	Suitable	Suitable
	Small parts	-	Suitable	Suitable	Suitable	Suitable	Suitable
Coats	Chest-piece	-	Suitable	Suitable	Suitable	Suitable	Suitable
	Front	-	Suitable	Suitable	Suitable	Suitable	Suitable
	Small parts	-	Suitable	Suitable	Suitable	Suitable	Suitable
Trousers	Small parts	-	Suitable	Suitable	Suitable	Suitable	Suitable
Dresses	Small parts	-	Suitable	Suitable	Suitable	Suitable	Suitable
Blouses	Small parts	Suitable	Suitable	Suitable	Suitable	-	-
Shirts	Small parts	Suitable	-	-	Suitable	-	-
Pyjamas	Small parts	Suitable	-	Suitable	Suitable	Suitable	-

Table 3: Men's-wear Applications

Article	Resin / Area	Low-density Polyethylene A	Medium-/High-density Polyethylene B	Polyamides C	PVC D	Cellulose Acetate E	Phenolics F
Jackets	Chest-piece	-	-	Suitable	Suitable	Suitable	Suitable
	Front	-	Suitable	Suitable	Suitable	-	-
	Small parts	-	-	Suitable	Suitable	Suitable	Suitable
Overcoats	Chest-piece	-	-	Suitable	Suitable	Suitable	Suitable
	Front	-	-	Suitable	Suitable	-	-
	Small parts	-	-	Suitable	Suitable	Suitable	Suitable
Trousers	Small parts	-	-	Suitable	Suitable	Suitable	Suitable
Knitwear	Front	-	-	Suitable	Suitable	-	-
	Small parts	-	-	Suitable	Suitable	Suitable	Suitable
Shirts	Front	Suitable	Suitable	-	Suitable	-	-
	Small parts	Suitable	Suitable	-	Suitable	-	-

Table 4: Ladies'-wear Applications

Article	Area \ Resin	Low-density Polyethylene A	Medium-/High-density Polyethylene B	Polyamides C	PVC D	Cellulose Acetate E	Phenolics F
Suits	Chest-piece	-	-	Suitable	Suitable	Suitable	Suitable
	Front	-	-	Suitable	Suitable	-	-
	Small parts	-	-	Suitable	Suitable	Suitable	Suitable
Dresses	Small parts	-	-	Suitable	Suitable	Suitable	Suitable
Blouses	Small parts	Suitable	Suitable	Suitable	Suitable	-	-
Overcoats	Chest-piece	-	-	Suitable	Suitable	Suitable	Suitable
	Front	-	-	Suitable	Suitable	-	-
	Small parts	-	-	Suitable	Suitable	Suitable	Suitable
Knitwear	Small parts	-	-	Suitable	Suitable	Suitable	Suitable
Trousers & Skirts	Small parts	-	-	Suitable	Suitable	Suitable	Suitable

Table 5 : Other Applications

Article	Resin / Area	Polyamides	PVC	Poly(vinyl acetate)
Rainwear	Front Small parts	Suitable Suitable	Suitable Suitable	- -
Leather and Fur	Front Small parts	Suitable Suitable	- -	Suitable Suitable

The development and improvement of resins are an on-going process, with new products being continually added to the existing wide range of options. Thus the garment producer has a growing choice from which to 'fine-tune' the fusibles used for specific cloths and effects.

4. COATING SYSTEMS

Coating is the process whereby the thermoplastic resin is deposited and secured onto the substrate material. There are many coating methods in use, and they all involve the application of carefully selected particle-size fractions of the particular resin being used.

The resin powders are produced by ambient (room-temperature) and freeze-grinding techniques. After grinding, sophisticated sieving techniques are used to separate the various fractions, which are designated by microns or micrometers (one-millionth of a metre, or µm). The micron ranges for the most widely employed coating methods are:

Method	**Micron Range**
Paste coating	Up to 80 µm
Dry-dot printing	From 80 to 200 µm
Scatter coating	From 150 to 350 µm

Very fine powders for paste coatings are sometimes produced by precipitation from solvents. This process, while producing a more uniformly sized particle, is far more expensive than ambient or freeze-grinding techniques.

Some of the main coating methods are as follows.

(a) *Scatter Coating* One of the earliest methods used for applying resin to the base cloth was by means of a vibratory feeder. This has since been replaced by scattering heads, which electronically control the amount and evenness of the coating that falls onto the moving substrate. After scattering, the coated base cloth enters an oven, where the resin softens and is pressed onto the base cloth. This is still the cheapest method of coating the base cloth, but the end-result is not as uniform and flexible as that of printed coatings (*Fig 15A*).

(b) *Dry-dot Printing* In this process, the resin powder is applied to a roller engraved with small indentations, which hold the powder. The amount and shape of the indentation are predetermined according to the required concentration of resin. These indentations can range from 2 to 9 dots per square centimetre.

The dot-printing process starts by passing the base cloth over a heated roller and then against the engraved roller, which contains the resin in its indentations. The heat transferred from the cloth partially melts the resin, which then adheres to the base cloth and can be withdrawn from

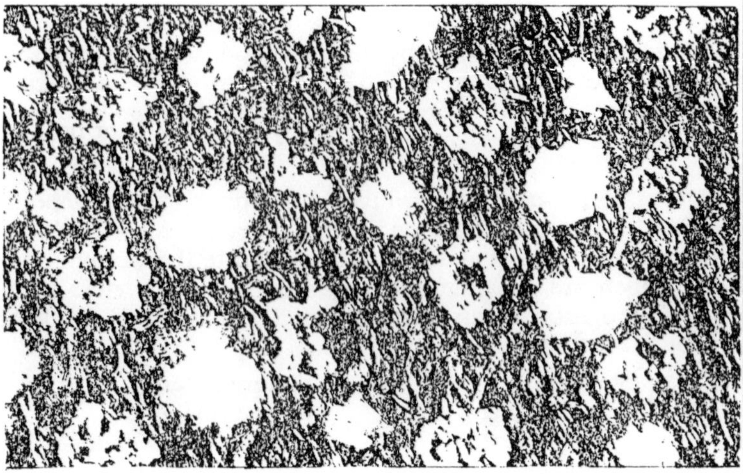

Fig 15A : Scatter coating (spray-sintered)

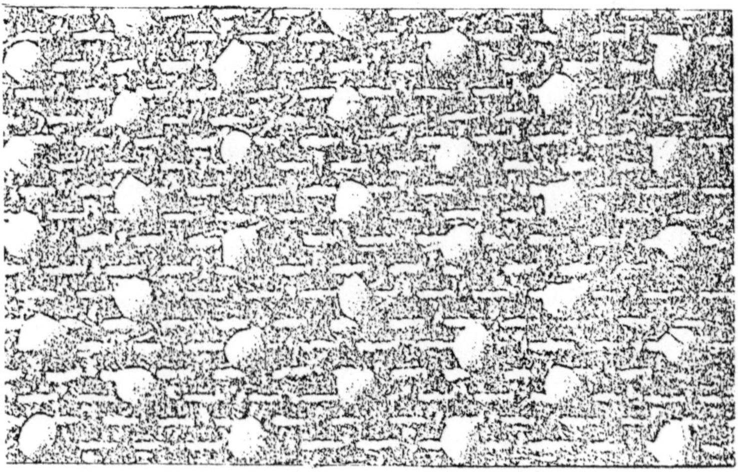

Fig 15B : Dry-dot printing

the indentations in the form of dots (*Fig 15B*).

As a general rule, lightweight fabrics require fusibles with small dots, whereas heavyweight fabrics require large dots. This is because there has to be sufficient resin to penetrate the fibre surface of the top-cloth to the depth required in order to produce a satisfactory bond strength. Dot coatings can be manipulated to have little or no effect on the properties built into the base cloth in terms of handle and crease-recovery. It is generally acknowledged that dot-printed fusibles produce a very superior type of final laminate.

(c) *Paste Coating:* After fine grinding, the resin powder is blended with water and other wetting agents to form a smooth paste. This solution is then printed in dot form onto the base cloth by means of a cylindrical printing screen. After printing, heat is applied in order to remove the water and to coalesce the printed dots into solid-resin points.

This coating method produces a precisely shaped dot and, in particular, is used to produce the finer-print dot required for the top collars of shirts. A paste of plasticized PVC can also be applied by this method.

(d) *Preformed:* The resin is heat-processed to form a net (*Fig. 16*), which is then laminated onto the base cloth by means of heat and pressure. During the heating, the net melts and forms a minute dot pattern on the substrate. This type of coating is widely used for the top collars of shirts.

(e) *Extrusion Laminating:* Thin films of molten polyethylene are laminated directly from an extruder onto an interlining base fabric to form a continuous heat-sensitive adhesive film. This type of interlining produces a stiff impervious finish and, over the years, has lost its popularity as a fusible for shirt collars.

(f) *Hot-melt Coating:* This method is used to produce continuously plasticized polyvinyl coatings on interlinings intended for leather goods and millinery. A marked disadvantage of this method is that the resin is held at its melt temperature for a considerable period of time, which causes discoloration and degradation.

(g) *Emulsion Coating:* This was one of the first methods used for thickened poly(vinyl acetate) emulsions. The emulsion is usually spread onto the base cloth by means of a blade or, for double-sided coatings, the base cloth is dipped into the emulsion and the excess squeezed out by passing the base cloth through rollers. These types of fusible are still used for leather and fur.

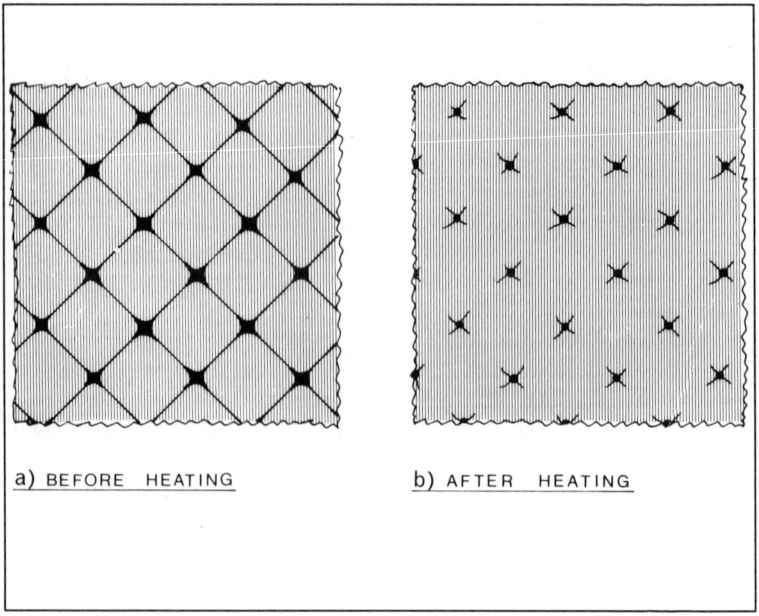

Fig 16 : Preformed resin net

(h) *Spray Coating:* From time to time, various resins have been applied to the base cloth by means of spraying. In the main, solutions of alcohol-soluble polyamides have been sprayed onto nonwoven fabrics in order to produce low-melt, dry-cleanable coatings.

The consideration of a coating system is basically concerned with the required degrees of flexibility and uniformity. These factors have to be evaluated in conjunction with all the other requirements and parameters that influence the selection of a particular fusible.

5. SELECTION OF FUSIBLES

The selection of fusible interlinings is governed by many objective factors, but, in the final analysis, choice is usually determined by the cost/performance relationship. Owing to the wide range of fusibles available, it is advisable to pursue a systematic selection procedure, which would involve the consideration of the following factors.

(1) It should never be assumed that fusing is possible in all cases without checking these points.
   (a) Will the top-cloth withstand the fusing conditions without excessive shrinkage, glazing, and changing colour?
   (b) Does the construction of the top-cloth include continuous-filament yarns or incorporate a special resin finish (e.g., silicon) that could adversely affect the durability of the fused product?
   (c) Does the top-cloth in question have an open construction that will allow 'strike-through' of the resin?
   (d) Does the top-cloth have a pattern or surface that could be damaged by the pressure used during fusing?

(2) The next stage is to consider whether a woven, knitted, or nonwoven type would serve the purpose. There is no simple answer to this question, since a great deal depends on:
   (a) particular garment style;
   (b) top-cloth type;
   (c) required appearance, handle, and drape characteristics.

   Since nonwoven interlinings are generally less expensive than other types, it is advisable first to assess their suitability. It should be remembered that different parts of the same garment could require different grades of fusibles. The evaluation of substrate material should therefore also take into account the function of the fusible in relation to the area and/or component being fused.

(3) There are the considerations of resin type and coating method, which are basically concerned with the question of whether the garment is intended to be dry-cleaned only, washed only, or both. If so, does the fusible require any special conditions of dry-cleaning or washing? Generally, the method used to clean a garment is determined by the type and composition of the top-cloth, and all the trimmings, such as linings and fusibles, are matched to these conditions.

(4) Another factor influencing the selection of fusibles is the type and quality of the fusing equipment available in the

factory. Fusing requires the application of heat and pressure, and a very important consideration is the method employed to generate the heat. The main issue is whether to use steam heat or dry heat.

(a) Steam Heat: A well-known adage in the clothing industry regarding the use of steam for fusing is: 'If steam produced the marriage, then it can also cause the divorce'. When the fusible resin is originally activated by steam heat, the same thing could happen in the subsequent pressing operations where steam is used. Since some areas of a garment are subjected to two or three pressing operations, the handle or lamination quality or both can undergo serious degradation. To avoid delamination positively during pressing or pressing after dry-cleaning or laundering, the initial fusing should occur at a temperature higher than that of steam-pressing.

(b) Dry Heat: Fusing is not a finishing operation but rather a permanent part of garment assembly. As such, this assembly must be performed on equipment that precludes the possibility of later delamination. The most important characteristics for permanent fusing are therefore temperature setting and control. Dry-heat fusing provides the degree of control necessary to achieve the correct bond at the beginning of garment assembly that will not be affected by further steam-pressing operations. Steam heat cannot guarantee the same permanence of bond as that achieved by the use of dry heat.

The use of dry-heat fusing can sometimes cause a change in the colour of the top-cloth. This change is usually temporary and is due to the fact that the atmospheric moisture is removed during fusing. After a short time, when normal atmospheric moisture is again present in the fabric, the original colour shade will return. Steam-pressing also aids this process by adding moisture to the top-cloth.

(5) Cost This is a very important factor, but it should not necessarily be the sole criterion for selecting a particular fusible. Price itself means very little without balancing it against the following considerations:

- *(a)* the desired performance of the fusible;
- *(b)* compatibility with existing fusing equipment; and
- *(c)* the technical services and back-up provided by the supplier of the fusible.

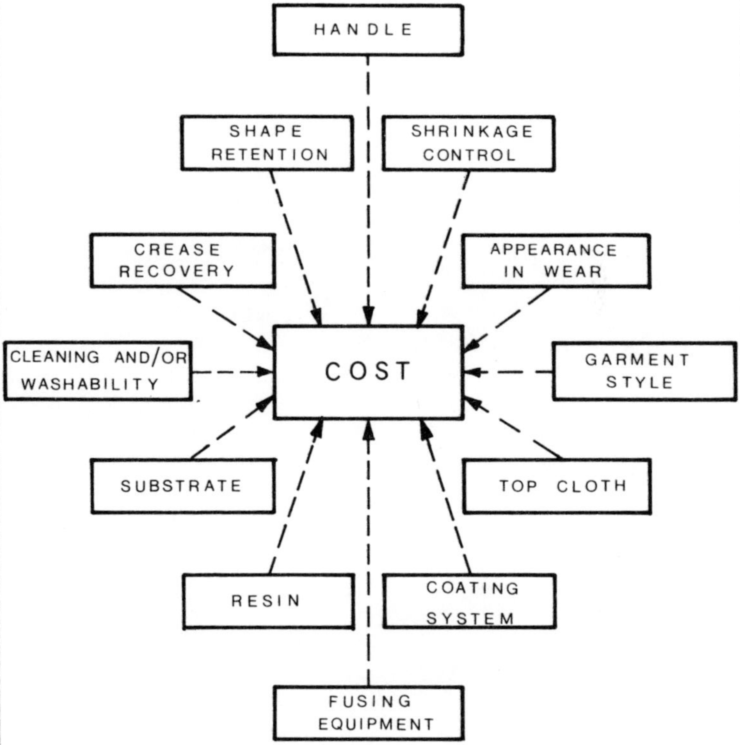

Fig 17 : Factors influencing the choice of fusibles

> There are obviously other commercial factors to take into account, but the fusible that is finally selected should be that which produces the desired garment at the lowest cost -- a cost/performance decision.

A summary of the factors concerned with the selection of fusibles is shown in *Fig 17*.

Finally, an all-purpose fusible that suits every fabric and can be fused at any temperature and with any pressure has yet to be invented.

6: THE METHOD AND COMPONENTS OF FUSING

The technology of the fusing process demands a specific combination of the following parameters:

(a) the fusing method;
(b) the fusing machine; and
(c) the components of the fusing.

(1) Fusing Method: The relationship between the top-cloth and the fusible influences the fusing conditions, and it is essential that the press can achieve the correct combination for the particular method employed. The following are the most widely used fusing methods, and the descriptions refer to the assembly of the cloth and fusible components prior to fusing (*Fig 18*).

(a) Single Fusing: The fusible is positioned on top of the cloth (*Fig 19*).
(b) Reverse Fusing: The cloth is positioned on top of the fusible.
(c) Sandwich Fusing: Here, two components are fused in one operation, typically with the cloth on the outside and the two fusibles in the middle.

Fig 18 : Fusing methods

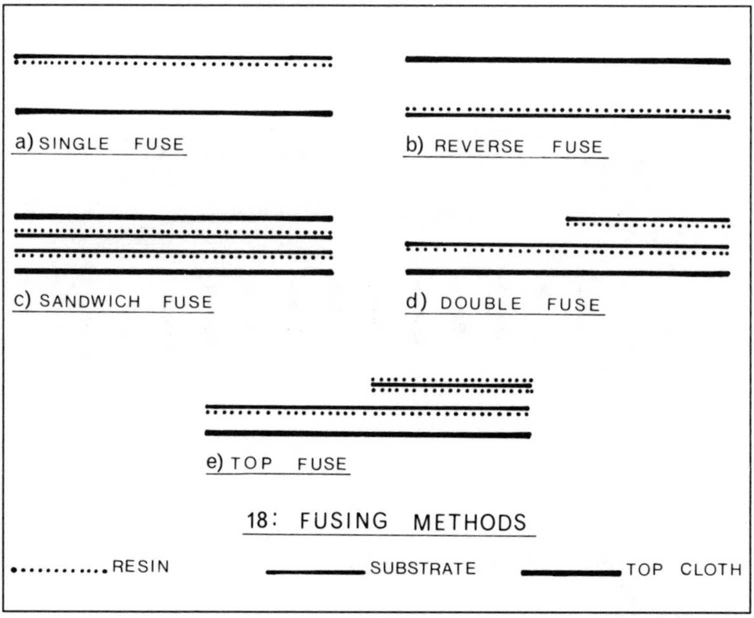

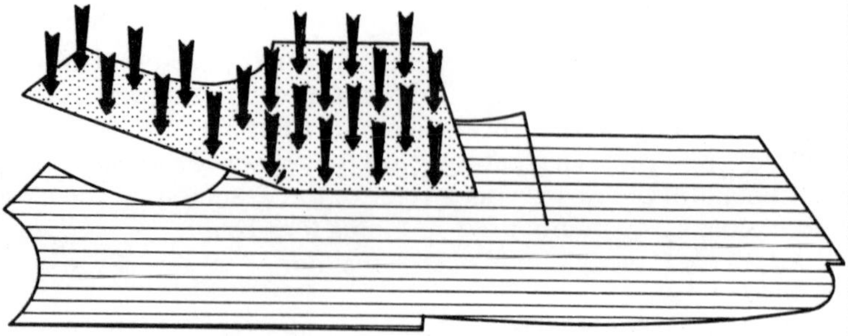

Fig 19 : Heat applied to fusibles

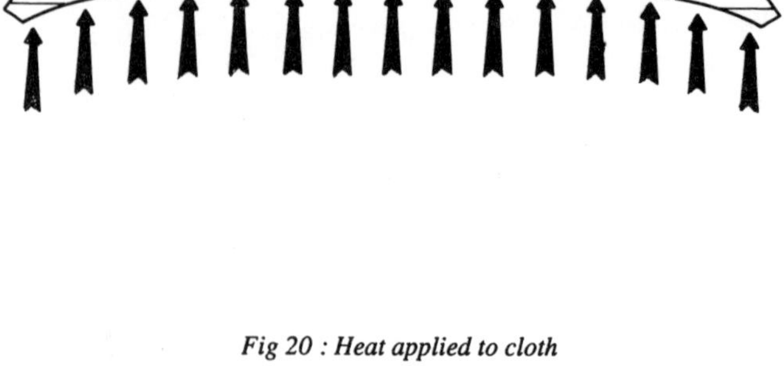

Fig 20 : Heat applied to cloth

(d) Double Fusing: This consists in fusing two fusibles, positioned one on top of the other, to the top-cloth in one operation, for example, fusing the front and chest-piece fusibles to the forepart of a jacket or coat.

(e) Top Fusing: The fusible components are positioned on top of the cloth, and heat is applied directly to the top-cloth. This method is used almost exclusively for shirt collars where single- and double-sided fusibles are part of the laminate construction (*Fig 20*).

(2) Fusing Machines The two basic types of machinery used for fusing are as follows.

(a) Flat-bed Press : This is a static method whereby the assembly is fed or positioned onto the bottom plate and the head, or top plate, is closed onto the assembly. At the end of the fusing cycle, the assemblies are cooled and then removed (*Fig 21A*).

(b) Continuous Operation: This type of machine has an endless-conveyor system for transporting the assemblies successively through the heating, pressure, and cooling stages (*Fig 21B*). Fusing machinery and equipment will be examined in more detail in the next chapter.

(3) The Components of Fusing Irrespective of the fusing method and type of machinery employed, fusing is controlled by four processing components:

(a) temperature,
(b) time,
(c) pressure, and
(d) cooling,

and they must be accurately combined in order to achieve optimum results.

(a) Temperature

There is a limited range of temperatures that are effective for each fusible resin. With too low a temperature, there is an inadequate development of the adhesive properties. Too high a temperature causes the resin to become too fluid, which could produce a possible strike-back or strike-through of the resin. In general, resin-melt temperatures can range from 130 to 160°C, and their optimum performance will normally occur within ±7°C of the specified temperature.

This situation requires a heating source and system that can ensure the maintenance of the selected resin-melt temperature for fusing cycles of varying times. Practically, this means that the heating source must have a very high capacity coupled with excellent conductivity in order to ensure an even distribution of the

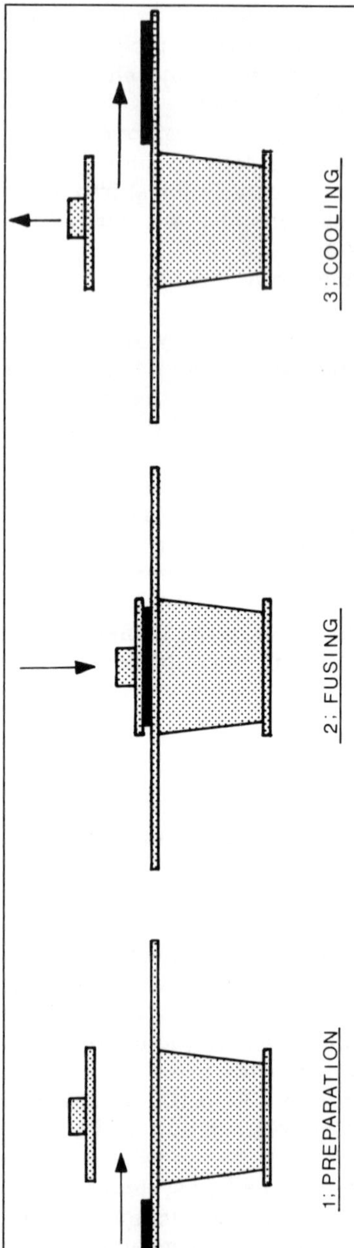

Fig 21A : Flat-bed press-: Schematic

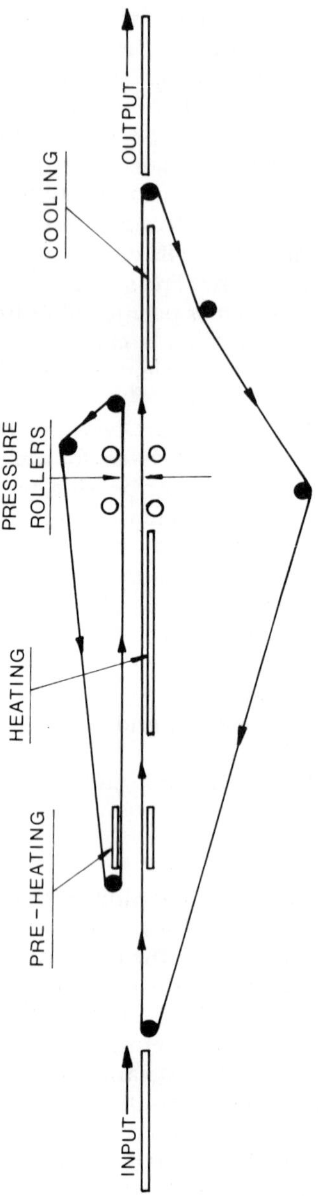

Fig 21B : Continuous cperation-: schematic

temperature over the entire heating area. All modern fusing machines are fitted with temperature-control thermostats, and, in practice, they should control the applied temperature within a maximum range of ±5°C. The thermostat should be precise and react immediately to any slight variations in the fusing-area temperature. The two graphs (after R.A. Burfitt) in *Fig 22* illustrate the difference in reaction time between slow and quick thermostats. The graphs show that the thermostat with a faster reaction produced peaks of temperature within the fusing time, whereas the reaction time of the slower (cheaper) model was not short enough to achieve an applied temperature within the allotted fusing-time cycle.

Most thermostats fitted to fusing machines show the indicated temperature of the heat source and not the actual applied temperature in the resin-heating zone. The difference between these two temperatures is due to:

(a) covers over the heat source, such as conveyor belts, protective frames, etc.;
(b) the pressure-support system;
(c) the residual moisture content of the fabrics being fused;
(d) the insulating properties of the cloth and fusible;
(e) machine environment, i.e., air agitation, ambient temperature, etc.

It is therefore essential to be able to measure the applied temperature, and this is usually done by using thermal papers or electro-mechanical calorimeters. The readings thus obtained are used, when necessary, to calibrate the indicated temperature settings.

For practical purposes, thermostat dials must be easily seen and read and located where they can be easily operated. The control itself should be capable of being locked at a particular calibration so that any accidental movement of the setting is avoided. The calibration and control of the applied temperature are central to the quality of the fusing process. *Fig 23* shows the control panel of a modern continuous-fusing press.

(b) Time

The only time element that is of any value in fusing is that time when the assemblies are actually being heated. The function of this time element is to compensate for the delay in the transmission of heat from the source to the fusible agent (resin). This time component relates to:

(1) the time when the two plates of the press actually meet with the correct specific pressure, or
(2) the time for which the assemblies are actually in the heating

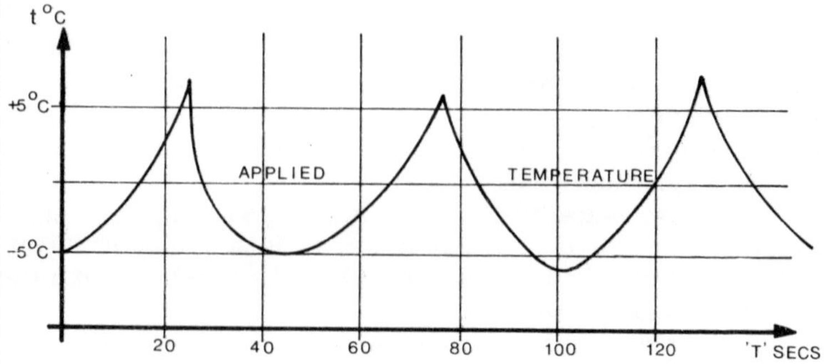

Fig 22a : Slow-reacting thermostat

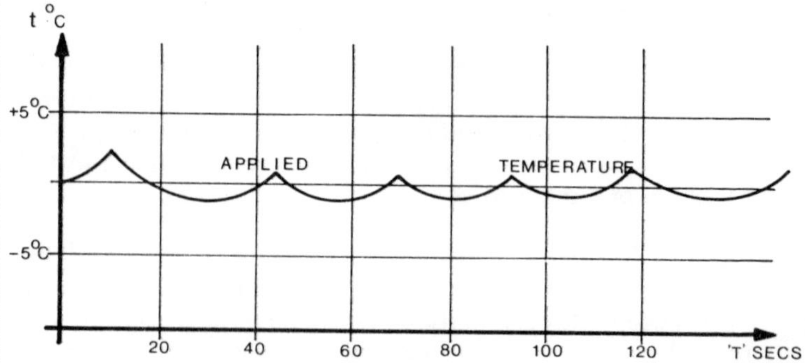

Fig 22b : Quick-reacting thermostat

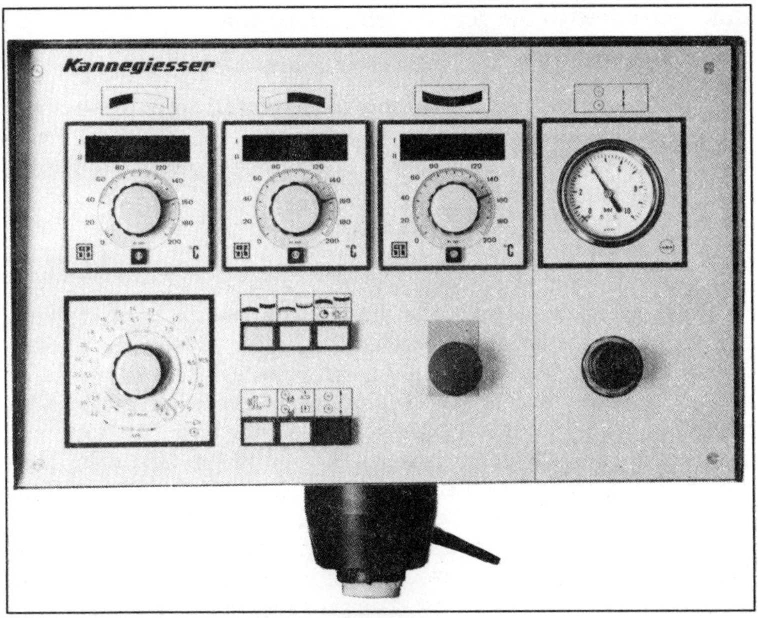

Fig 23 : Control panel of continuous-fusing press

zone if a continuous machine is employed.

The actual fusing time can be controlled by the use of electro-mechanical timing devices or programmed card units. Generally, a flat-bed press has a mechanical timer that runs to the zero point and then signals an impulse, which activates the mechanism for opening the press. With continuous fusing, it is possible to relate conveyor speeds to the actual time in the heating zone. The accuracy of the timing controls can be verified by comparison with a stopwatch or by measuring the conveyor-belt speed with a tachometer.

The particular time cycle to be used for any particular fusible is based on:

(a) whether the interlining has a high- or low-melt resin system;

(b) if a light or heavy substrate is being used;

(c) the nature of the top-cloth, the amount of moisture present, and whether the cloth has been siliconized;

(d) whether large or small areas have to be fused;

(e) the type of fusing equipment, i.e., continuous, flat-bed, steam-press, or hand-iron.

Most manufacturers of fusibles supply data sheets detailing

the time cycle for a particular fusible, and it should be noted that this time refers to the actual fusing stage only.

(c) Pressure

With the resins in current use, it is necessary to apply pressure to the parts being fused in order to ensure:

(a) that an intimate contact is effected between the top-cloth and fusible;

(b) that the heat transfer is adequate;

(c) that there is a controlled and even penetration of the resin into the fibres of the top-cloth.

Pressure is usually applied via platens, bucks, or nip rollers, depending on the type of machine being used. The energy used to create pressure can emanate from pneumatic, hydraulic, or manual sources and is transferred to the pressure surfaces by means of cylinders or mechanical linkages or both. Suppliers of fusibles will generally indicate the specific or applied pressure at which the optimum bond strength is obtained for a particular fusible. This figure is not the same as the air supply to the machine but is the stated pressure applied to the pressure plates or rollers. Too much or too little pressure can result in inadequate bond strength and other problems. The pressure component of fusing is no less important than the temperature component. The pressure system of a continuous press is shown in *Fig 24*.

(d) Cooling

The effect of enforced cooling on fused assemblies is to achieve a consolidated bond, which enables the assembly to be handled immediately after cooling is completed. Enforced cooling can be brought about by:

(a) water-cooled plates;

(b) forced-air circulation;

(c) vacuum;

and the cooling method must be capable of reducing the latent temperature of the fused assemblies to approximately 80°F without affecting the productivity of the fusing machine. The assembly should be cooled when flat and in a horizontal plane and the cooling applied so as not to distort the assemblies. Enforced cooling ensures that higher levels of productivity are achieved than if the fused assemblies are left to cool naturally.

To sum up, fusing is not a difficult or complicated technology, and the optimum results can be achieved by the accurate and continual control of the four processing components.

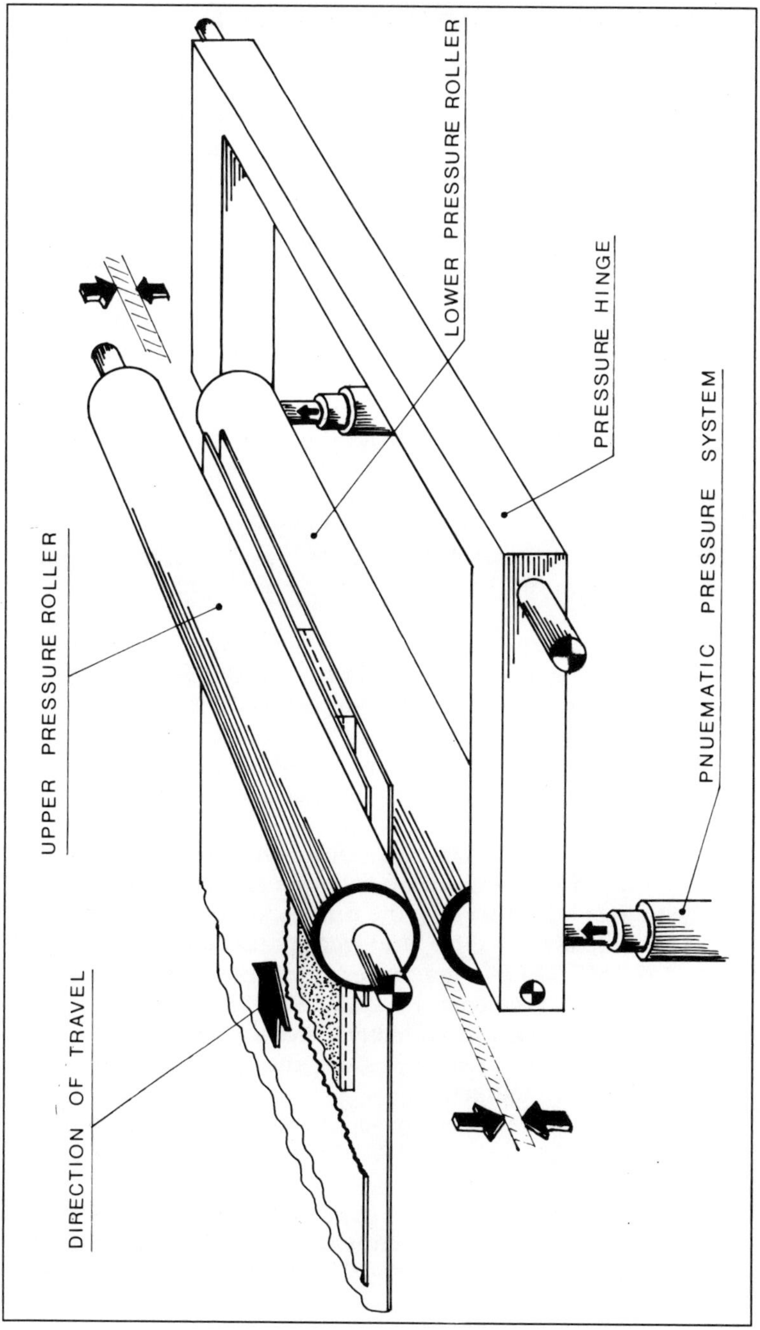

Fig 24 : Pressure system of continuous press

7. MACHINERY AND EQUIPMENT

(1) General Requirements: The mechanical medium required for fusing is the fusing press, and investments in this type of equipment must be carefully appraised in order to make rational decisions. The wide range of fusing presses available today has created a problem of choice as to which system offers the best solution to the requirements of a particular factory. In the final analysis, the decision to purchase a fusing press is based on the required applications of the machine, and each factory has its own considerations regarding the advantages and limitations of a specific fusing system.

A handbook of this nature cannot make a detailed evaluation of every available system but can only examine the main technical principles involved. This examination commences with some pertinent questions that should be asked in contemplating the purchase and installation of a fusing press.

(a) Method: Is the machine capable of providing the combinations required by the fusing methods employed in the factory?

(b) Capacity: What production of large and small parts is required? This question must be related to the loading time, fusing cycle, and take-off time, especially in the case of flat-bed presses.

(c) Loading: How and where are the units for fusing assembled, and how are they fed into the machine? Directly via trays or by conveyor belt?

(d) Instrumentation: Are the instruments of measurement that are fitted to the machine accurate and reliable? Can the settings for controlling the functions be adjusted easily?

(e) Cooling: Is it effective, and what time element is involved?

(f) Take-off: Whether manual or automatic? Can a stacker for small and large parts be attached to the press?

(g) Cleaning: Can the protective covers or conveyor belts or both be easily cleaned? Does the machine have a self-cleaning mechanism?

(h) Protection: Are there provisions against heat emission, loss of power, and heat loss? What are the working-environment conditions?

(i) Safety: Are there devices to prevent accidents and to stop the entry of wrongly positioned components? Is the machine fitted with an emergency stop?

(j) Space and Location: Where will the machine be located, and

what space is required? Does the chosen area have the strength to support the imposed load of the machine and ancillary equipment?

(k) Assembly: How will the machine be delivered — or assembled in sections? Are the doorways and passages large enough to permit the movement required in order to position the machine in the chosen location? Would structural alterations be required, and, if so, to what extent?

(l) Installation: What energy sources are required — electricity, compressed air, vacuum, water, etc.? Are the existing sources sufficient, or will additional installations be necessary?

(m) Service: What technical help can the supplier provide, and what is the availability of spare parts? What types and at which levels should spare parts be carried by the factory in order to minimize machine downtime?

(n) Maintenance: Can general and preventive maintenance be carried out by factory personnel? Is any special training required? Where will the training take place, and who will provide it?

(o) Cost: What is the total installed cost of the machine? This could be considerably higher than the price of the machine itself.

(p) Pay-back: What criteria will be used to calculate pay-back time? What pay-back time would be acceptable in relation to the total investment required?

(q) Investment: Can the factory cover the investment required, or must the capital be raised from outside sources? If self-financed, what is the cost of diverting the capital to the investment?

These questions and many others have to be answered in order to establish the cost/benefit relationships of a particular fusing system. It should be remembered that time is money; the amount saved by the purchase of a cheap press could therefore be far outweighed by possible losses due to machine failure or quality problems caused by faulty fusing. *Caveat emptor.*

(2) Fusing Presses: Presses basically fall into two categories, with different modes of operation. The assemblies are either fused on a flat bed in a static situation or conveyed continuously through a press in which they are heated, fused, and cooled without stopping. All presses can be described as flat-bed or continuous, irrespective of their size or complexity. Some of the equipment in general use is described in the following pages.

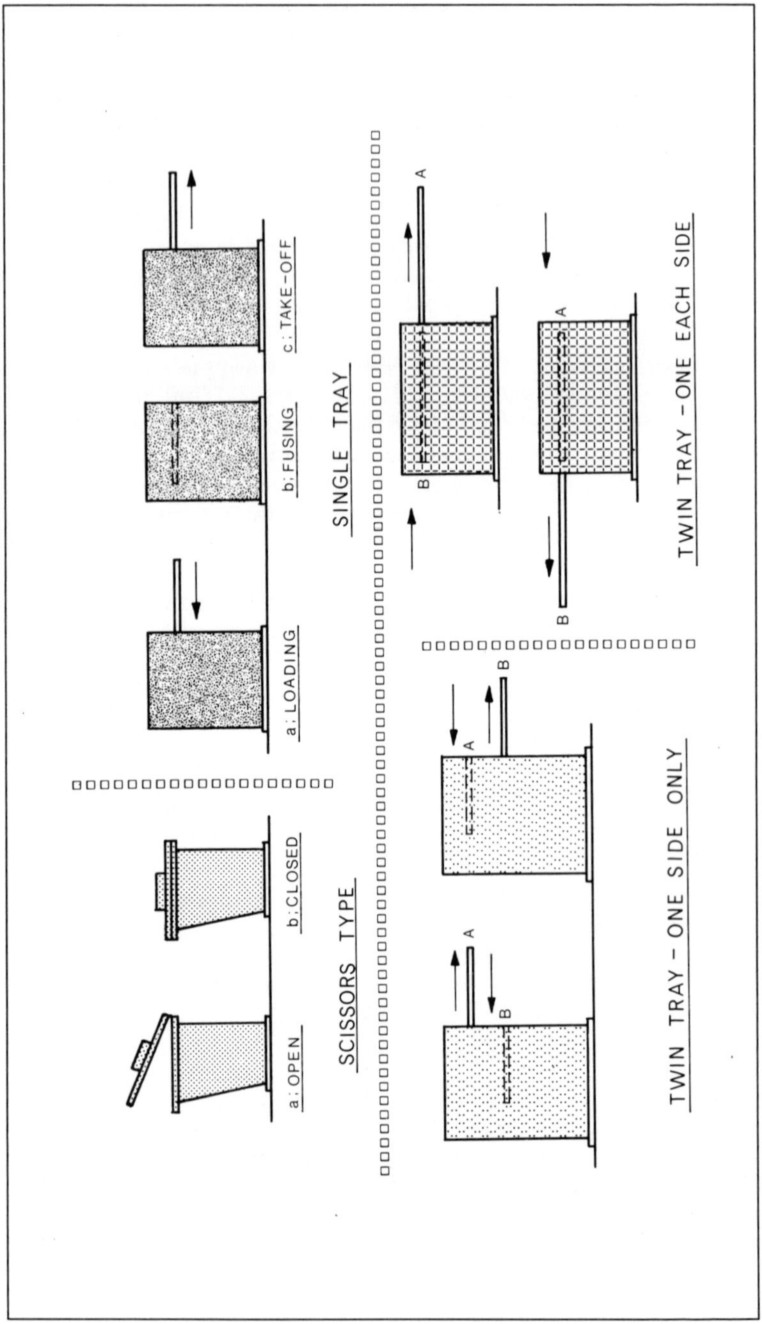

Fig 25 : Types of flat-bed press

(A) Flat-bed Presses (Fig 25)

(1) Steam Presses Steam presses with shaped bucks are not designed as fusing presses but have been, and still are, used for fusing. Whereas there are certain resins that can be used on steam presses, these machines have a number of serious limitations for general use, as follows.

(a) They are restricted in temperature, which means that it is impossible to reach the heat levels required by the majority of resins.

(b) Some presses are fitted with heated bucks that enable the machine to reach slightly higher temperatures. In either case, the press temperature is dependent on steam pressure, steam dryness, pipe insulation, and the number of presses drawing steam from the same line.

(c) The pressure applied over the full buck area is uneven, which restricts the machine's use to the fusing of small parts.

(d) There is no doubt that forcing steam through the laminate achieves a very rapid heat transfer. However, this could possibly cause excessive shrinkage of the top-cloth.

(e) The majority of pressing operations after fusing could cause the resin to replasticize. When the resin is in this condition, any movement, bending, or tension could partly weaken the bond and eventually cause delamination after dry-cleaning or washing.

(f) Most of the older types of steam press are not fitted with timers and programme controls, thus leaving the time element completely dependent on the operator.

Most suppliers of interlinings supply a data sheet, which states whether the particular interlining can be fused on a steam press and what are the conditions regarding component area, temperature, and time.

(2) Flat-bed Presses (Figures 26 and 27) These are available in a wide variety of systems, ranging from small table models to large, floor-standing models. There is also a wide choice of manual or mechanical loading and take-off systems that are usually built into the machine at the time of its manufacture. Basically, this class of press consists of padded top and bottom bucks, with heating elements in one or both of the bucks. The bottom buck

Fig 26 : Single-tray press

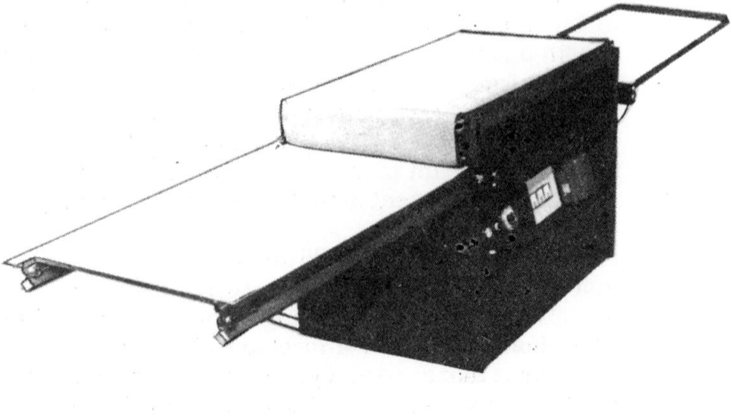

Fig 27 : Twin-tray press

is static, and the top buck can be moved vertically in order to open or close the press.

Some of the smaller models do not have a built-in feeding system, and the prepared assemblies have to be brought to the machine by hand, generally on a sheet of thick paper. After fusing, the sheet is removed, the parts are stripped, and the sheet is then prepared for another cycle. Some of the larger machines also use this method if the machinery has only a single tray. This method works as follows.

(a) The assemblies are either made up on the press tray or brought to the tray pre-prepared.

(b) The tray is then pushed or powered to its fusing position.

(c) When the fusing cycle is completed, the tray can be manually removed or automatically rejected.

Flat-bed presses with twin-tray systems are considerably more productive than those with single-tray systems because the time lag caused by waiting for the fusing cycle to end is considerably reduced. The main advantage of the twin-tray system is that, while one tray is being fused, the other can be stripped and loaded. Both the single- and twin-tray presses can be fitted with a device that automatically causes a non-stick fabric to roll over the unfused assemblies as the tray moves towards the fusing area. This fabric ensures that the fusibles are held in their correct positions *vis-à-vis* the top-cloth components.

Another type of flat-bed press is incorporated into a carousel by using three trays that successively move through the loading, fusing, and cooling stations. These machines are generally used for the 'top-fusing' of shirt and blouse collars (*Fig 28A,B*).

A machine used in conjunction with the 'top-fusing' method is the spot-welder. There are usually a number of different plies of fusibles that have to be assembled and fused to the collar. In order to ensure the exact positioning and relationship of the components, the parts are first laid into a special jig. Heated spots then press down onto the assembly and partly activate the thermoplastic resin. This combination of heat and pressure causes a small amount of adhesion between the plies that does not detract from the final fusing performance of the fusibles. Generally, this type of machine incorporates an automatic pick-up device for

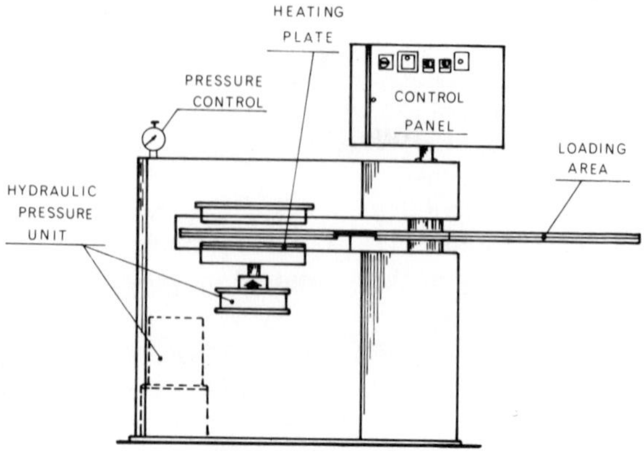

Fig 28A : Carousel press — side view

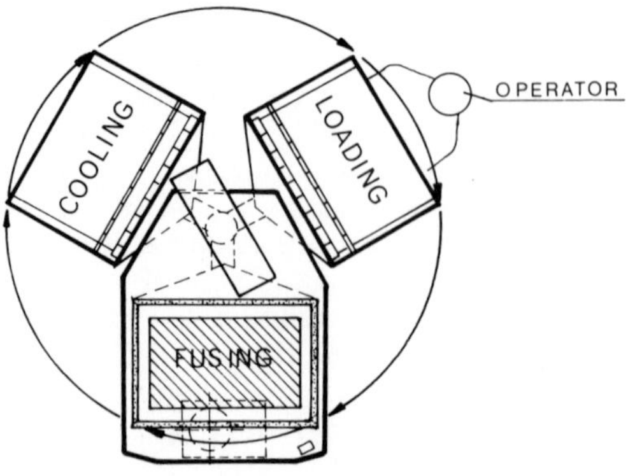

Fig 28B : Carousel press — plan view

Fig 29 : Spot-welding press

the various fusibles being used, and this substantially reduces the handling time involved in assembling the units (*Fig 29*).

A unique type of flat-bed press used for fusing bundles without disturbing the bundle formation incorporates two moving bucks. The top buck moves in the horizontal plane, to and from the operator, whereas the lower buck moves in the vertical plane. This action enables the operator to prepare the next ply assembly while the previous one is being fused. *Fig 30* shows the operating sequence, which is as follows.

(1) The bundle is positioned in the clamp, and the first component is lifted to the loading area on the top buck and the fusible positioned.

(2) The top buck then slides away from the operator, which causes the assembly to fall onto the exposed lower buck.

(3) The top buck moves back to the operator, and the bottom buck rises to exert heat and pressure on the assembly. The operator then prepares the next assembly in the loading area while the previous one is being fused.

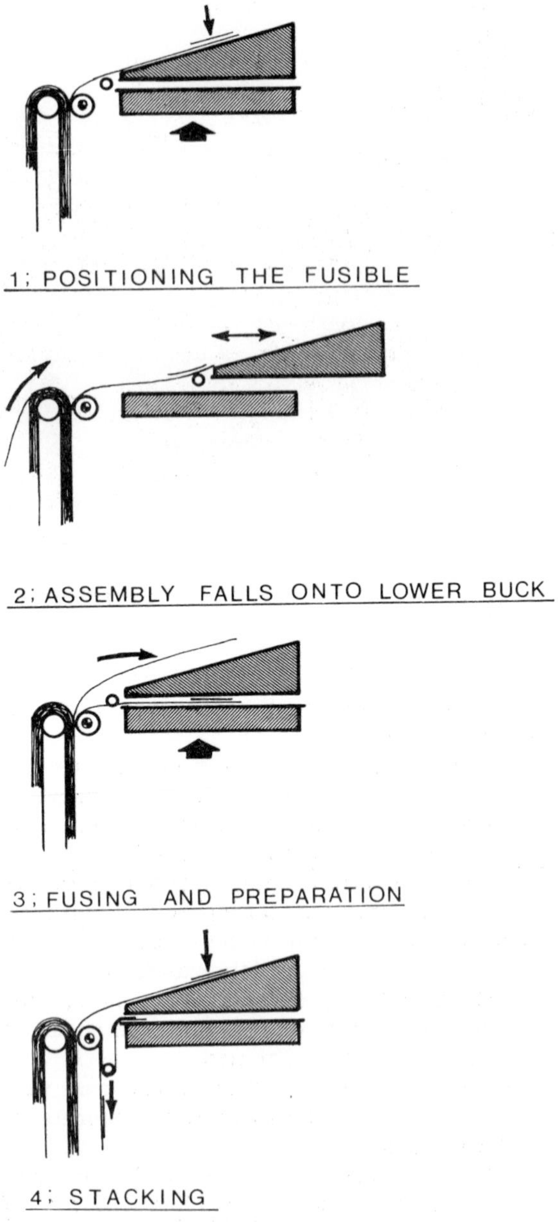

Fig 30 : Small-parts fusing press

(4) The lower buck is lowered, and the fused component is automatically restacked.

This machine is very efficient because the preparation and fusing operation are performed simultaneously.

Some general observations regarding flat-bed presses follow.

(a) *Pressure:* The even application of the correct pressure is essential for efficient fusing, and it is imperative to check that the press does not have low-pressure areas caused by the improper closing of the bucks. This could sometimes occur with scissors-type presses, which, because of their construction, have a tendency to leave a low-pressure area along the front edge of the press.

The operation of flat-bed presses requires the heating of a large mass of metal, and this heat could possibly cause the bucks to become warped. One solution to this problem is to keep the heating elements apart from the metal platen used to produce the pressure area. This type of construction is relatively costly and is found only in the more expensive type of presses.

(b) *Cladding:* All flat-bed presses have one padded buck that has the primary function of compensating for small irregularities in the metal platen. This padding must be capable of standing up to hard wear over a long period of time and retain its resistance to a temperature range of 100–200°C. It is generally considered that perforated silcone-rubber padding is the most suitable material for this purpose, although the initial cost is high. Felt types of material, which are sometimes used as an alternative, are cheaper, but, since they need frequent replacement, the final cost could be much higher. In addition, even when the felt type of cladding is new, it has the drawback of retaining moisture, which can affect temperature levels.

(c) *Suitability of Purpose:* The range of flat-bed presses is extremely large, and this is due to the fact that the clothing industry requires presses of different types and sizes for individual end-uses. There is no doubt that these presses are perfectly adequate if they are employed only for the end-uses for which they have been designed, which is usually the fusing of small-to-medium-sized areas. Problems could arise if their use is extended into fusing large areas for which the press does not have the temperature or pressure ranges and controls necessary for efficient fusing. The guiding principle should be, in all cases, the right machine for the job.

(3) Continuous-fusing Presses These fusing machines are all based on the same principle, whereby the unit for fusing is transported through the fusing processes by means of a powered conveyer belt. For practical purposes, the examination of these types of machine is concerned with the following factors.

(a) Conveyor Type: There are two systems in general use, and they are as follows.

(i) End-to-end Feed: The parts are conveyed from the loading area at one end of the machine, via the fusing area, to the take-off area at the opposite end of the machine. This system is used where production levels require continuous loading and feeding on the part of the operators.

(ii) Return Feed: This conveyor system is, in effect, two belts, one upper and one lower, where the lower belt is the actual conveyor of the cloth and fusibles. The function of the upper or cover belt is to ensure that both cloth and fusibles are held accurately in the correct positions while being transported through the fusing processes. Obviously, the speed of both belts must be synchronized in order to prevent the compression or extension of the materials *(Fig 31).*

(b) Cover Belts: The conveyor system is, in effect, two belts, one upper and one lower, where the lower belt is the actual conveyor of the cloth and fusibles. The function of the upper, or cover belt, is to ensure that both cloth and fusibles are held accurately in the correct positions while being transported through the fusing processes. Obviously, the speed of both belts must be synchronized in order to prevent the compression or extension of the materials.

(c) Heating: The heating systems generally used for continuous fusing presses are as follows.

Type 1: Heating Plates These consist of two heating surfaces, positioned apart, with one above the cover belt and one below the conveyor belt *(Fig 32).* The surfaces are slightly arched so as to intensify the heat diffusion. The lower heating surface is usually longer than the upper, and, in relation to the conveyor direction, the lower heat source is positioned after the higher one.

Both the panels can be separately controlled, which thus enables any temperature profile to be achieved.

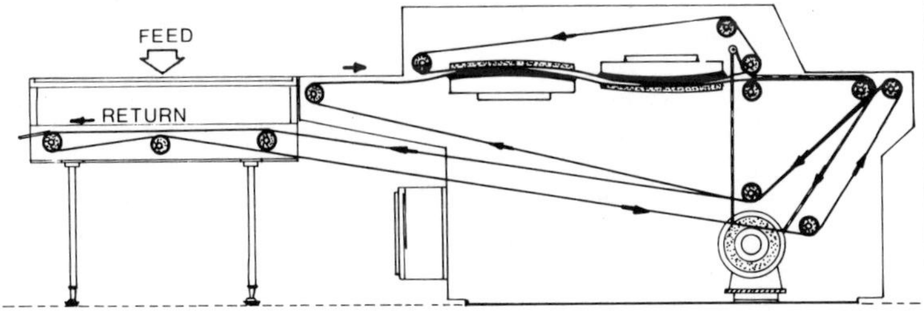

Fig 31 : Return-feed conveyor press

Fig 32 : Flat heating surfaces

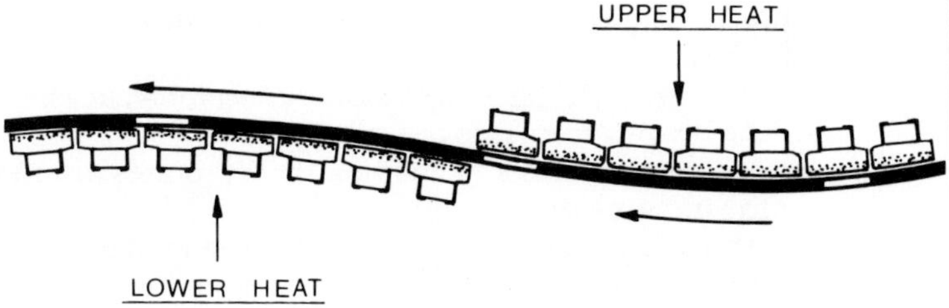

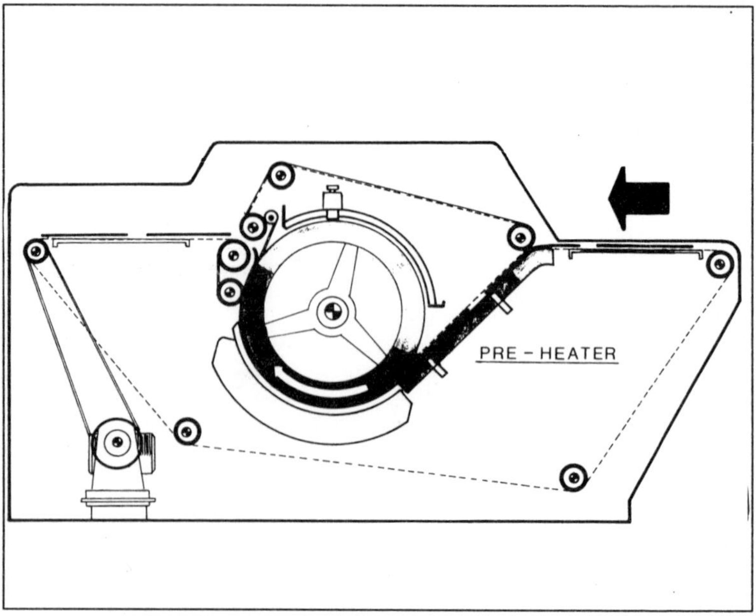

Fig 33 : Cylinder-heating system

For single or sandwich fusing, the temperatures of the upper and lower panels are usually the same. In fusing temperature-sensitive top-cloth, the temperature can be displaced towards the fusible by increasing the upper temperature and reducing the lower temperature. The reverse applies in fusing sensitive fusibles or heavy top–cloths or both.

Type 2: Cylinder Heating (Fig 33) The cylinder consists of two parts:

(i) the inner cylinder — a stationary assembly, in which the heating elements are mounted;

(ii) the outer cylinder or mantle, which rotates around the inner cylinder.

This principle ensures that the heat generated by the inner core is evenly distributed all over the cylinder mantle. In addition, since approximately half of the cylinder-mantle surface area is in the actual heating zone at any given moment, the remaining area regains the heat lost through transference.

These types of machine generally have a pre-heating device located under the lower conveyor belt, between the entry zone and cylinder. In fusing by the single-method, this heat source acts as pre-heater where the operating temperature is lower than that of the cylinder. For sandwich fusing, the operating temperatures of the pre-heater and cylinder are the same.

(d) *Pressure:* During the heating and plasticizing stages, only a slight pressure is exerted onto the components. Full pressure is applied to effect the final bonding after the components have been heated to the correct temperature. With continuous fusing, pressure is applied by passing the heated components through rubber-covered steel rollers positioned above and below the belts. The lower set of rollers are pressed against the upper set by means of compressed-air cylinders located at the ends of the lower roller. In addition, the diameter of the rollers and the hardness and thickness of the rubber covering also influence the final pressure exerted on the assemblies. These factors can be altered according to the types of top-cloth and fusible being used.

(e) *Cooling:* The fused parts are carried over the cooling zone by a separate conveyor band in order to prevent unnecessary cooling of the main conveyor belts. Cooling can be performed by the following means.

Fig 34 : Cooling systems

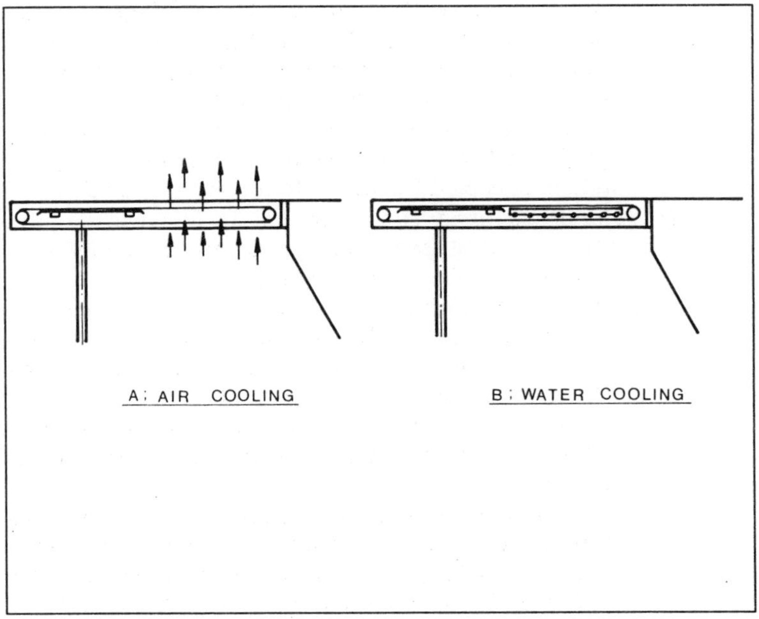

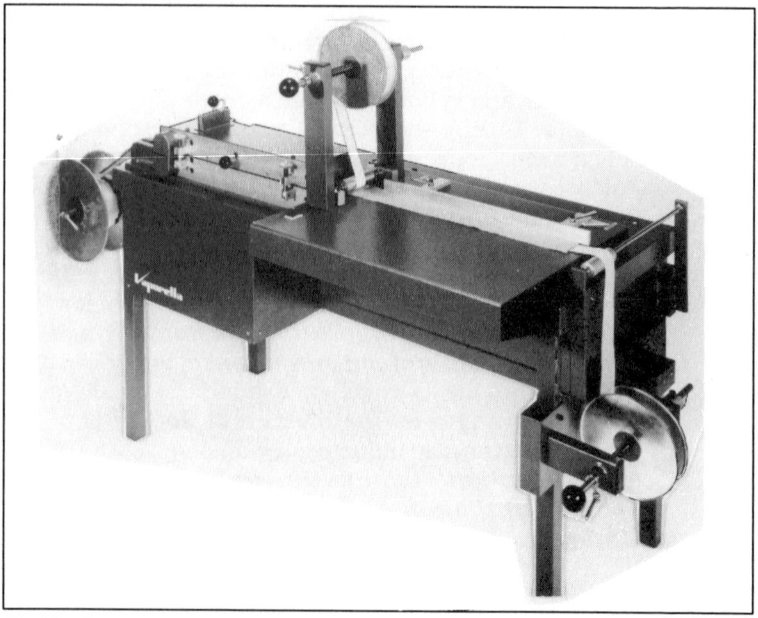

Fig 35 : Continuous-strip fusing

(i) *Air:* The air-cooling zone consists of a large number of thin bands, upon which the fused components lie. Cooling is affected by the circulation and convection of air around the components (*Fig 34A*).

(ii) *Water:* The fused assemblies are supported by a perforated belt positioned over an aluminium plate, which is cooled by water circulation (*Fig 34B*).

(f) *Take-off Systems:* The removal of the fused assemblies from the cooling area can be performed manually or by means of a mechanical stacker.

(i) *Manual System:* The requisite number of operators are positioned at the take-off position, and they remove the individual components and then stack them into bundles. This means that every assembly is handled individually and the time element is dependent on the feed-in rate at the opposite end of the machine.

(ii) *Mechanical System:* Various mechanical systems remove and stack the fused assemblies after they leave the cooling zone. In principle, these systems are segmented stackers that operate according to the feed rate of the assemblies and their position on the conveyor belt. The obvious advantage of using a mechanical system is that it requires handling of complete bundles and not of individual assemblies. (*This subject will be examined in more detail in the next chapter.*)

A variation of the conveyor-driven continuous-fusing press is a relatively small machine used mainly for fusing the waist bands of trousers and skirts (*Figures 35 and 36*).

These types of cloth components are generally fused to pre-cut fusible tapes in a roll form, and these tapes are sometimes perforated in order to provide a guide for sewing or folding. The tape, together with the lower conveyor belt, combines the functions of cloth carrier and transport medium during the fusing process. One or more reels of tape are mounted at one end of the machine, and the leading ends are fastened to a take-up wheel that revolves at the speed required for correct fusing. When the tape passes over the loading area, the operator positions the waist band on the moving tape, which travels through the heating, pressure, and cooling zones. The take-up spool is located after the cooling zone, and, when the assembly emerges from this area, it is wound up onto the spool. This

Fig 36 : Detail of continuous-strip fusing

method is also labour-saving because the fused assemblies for a particular bundle can be transported as one unit rather than as individual pieces. An additional use for this machine is the fusing of small, narrow parts, whose form is suitable for fusing onto a narrow, continuous tape.

(4) New Developments Two major innovations in fusing technology that have been developed in recent years are:

(A) high-frequency fusing, and

(B) direct stabilization.

Both of these technologies have revolutionized contemporary thinking regarding fusing, and there is no doubt that, within the next few years, these developments will become widely adopted in the clothing industry.

(A) High-frequency Fusing (Fig 37) : As the name suggests, this system utilizes high frequency to create heat, and, before examining this process, a short description of high frequency may be helpful.

Fabrics are non-conductive materials, and, with constant voltage, no electric current can be conducted through the material, i.e., from the top to the bottom of

Fig 37 : High-frequency fusing press

a stack of cut components. However, if an electric plate is positioned over the stack and another positioned under the stack and both plates are fed with high-frequency voltage, an electric field will be formed between the two plates. This field is continually commuted according to the applied frequency. The molecules of the materials between the plates form electric dipoles, i.e., they align themselves in the same manner as a compass needle in a magnetic field. As a result of the constant commutation of the electric field, the dipoles are forced to change their position constantly. Owing to the permanent friction losses, heat is produced mainly on the resin points themselves, and this is uniformly distributed over all the resin points in the whole electric field.

Since the resin coatings of interlinings have different dielectric properties from those of cloth, the resin is heated faster than the cloth. Thus the bonding can take place when the cloth is at 110°C rather than the 125-150°C required by other methods of heating. The fact that the top-cloth reaches a lower temperature ensures that most of the residual water is retained, which thus somewhat reduces the shrinkage that generally occurs during fusing at upper-level temperatures *(Fig 38)*.

The machine itself operates as follows.

(a) Preparation (Fig 39): The preparation of assemblies requires the interleaving of the cloth parts with the appropriate fusible, and the fusibles are laid in parts for sandwich fusing or individually for single fusing. Components are usually stacked to a height of about 10 cm and brought to the machine by means of a trolley or mechanical conveyor.

(b) Fusing Programme: This combines three parameters.

(i) Cloth Type: The fibre composition of the top-cloth has an influence on the time needed to attain the required heat. Generally, the heating time for synthetic fibres is less than that required for natural fibres.

(ii) Area: This is the percentage of the total area of the electrode plate that is utilized by the components to be fused. This is necessary in order to ensure that the same specific area pressure is applied to the area used.

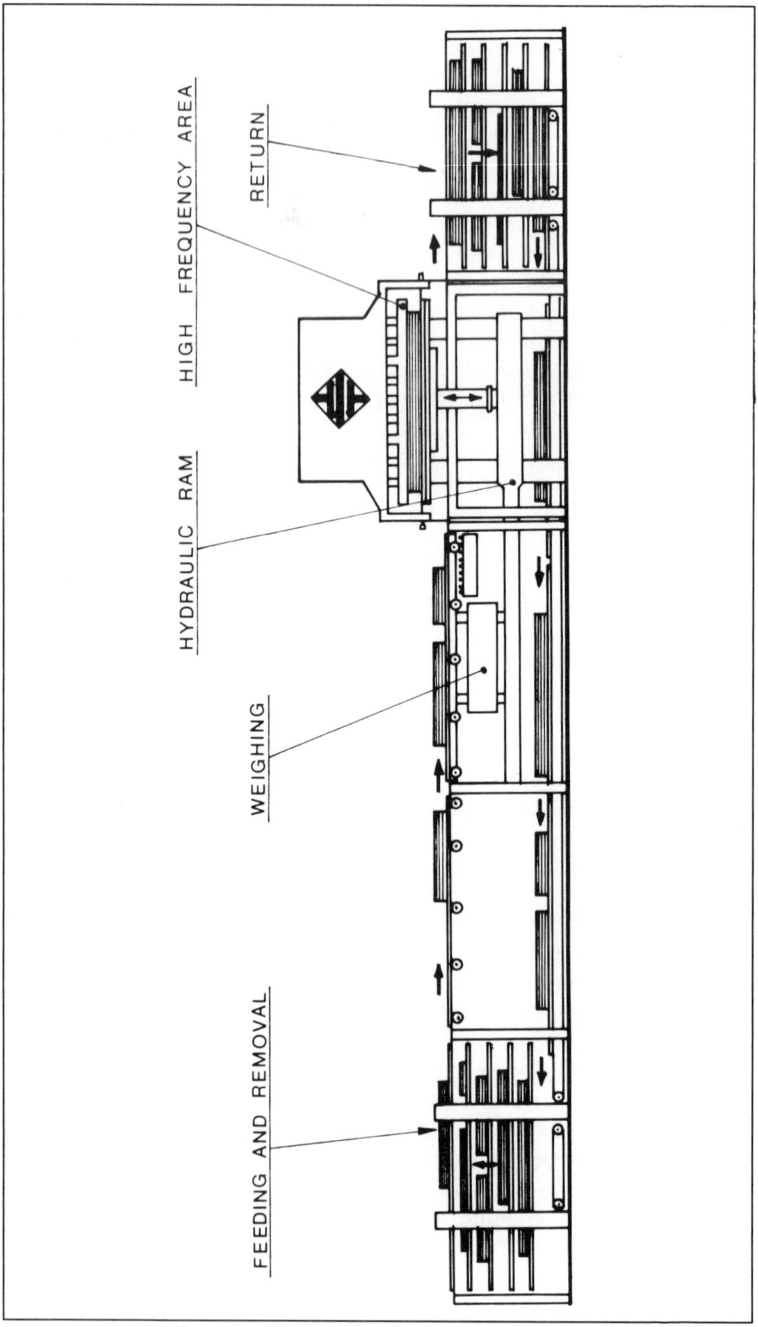

Fig 38 : High-frequency fusing press — work flow

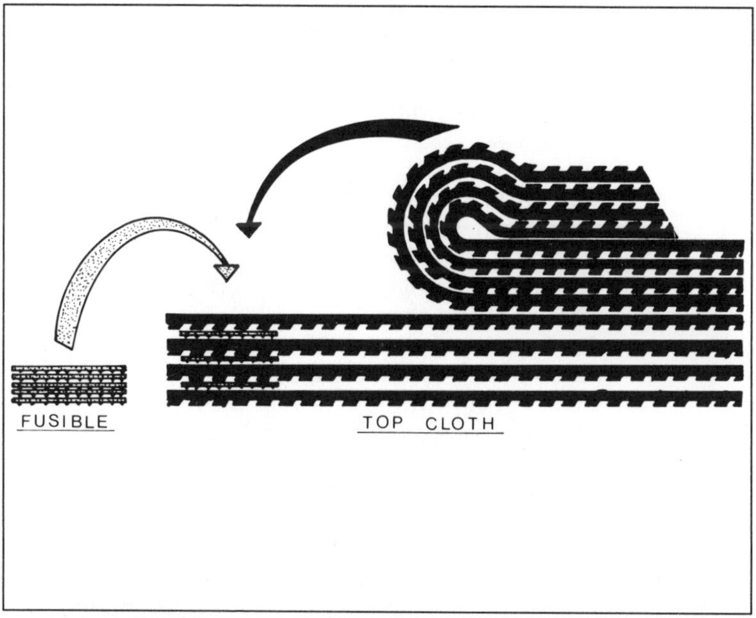

Fig 39 : Preparation for fusing by high frequency

(iii)Weight: The components on one plate are weighed in order to establish the amount of time required for heating the stacks.

(c) Operation: The stacks are positioned on fabric-covered frames that advance to the weighing station and are weighed. After setting the program, the frames advance to a heating-and-pressure zone, where, after the correct temperature has been reached, the press is closed by a hydraulic ram under the lower plate. When the bonding is completed, the trays emerge and are transported back to the take-off area by means of a conveyor belt running underneath the main working surface. During this travel, the fused components are cooled and can then be removed. The empty frames are returned to be prepared for loading.

According to the manufacturers and users of the high-frequency method, this offers a number of advantages over the systems currently being used.

(i) As the heating time can take up to 3 min instead of from 10 to 16 s, the 'heat shock' to the cloth is eliminated. (Heat shock is known to cause shrinkage and shine to the top-cloth.)

(ii) The resultant handle is softer because the resin dots retain their shape and do not flow into the top-cloth.

*(iii)*The peel strength of the bond is excellent and is retained even after being subjected to many dry-cleanings.

(iv) There are considerable savings in energy because heat is only generated during the actual heating period and not continuously.

(v) The total operation is very speedy because complete bundles are being handled or processed at each stage instead of one or two plies.

(vi) For safety reasons, all functions of the machine are automatically monitored, and any slight deviation is signalled optically and acoustically.

This system, with its novel technology, was initiated by a German clothing producer and developed jointly with the machine builder and a prominent manufacturer of fusible interlinings.

(B)Direct Stabilization: The unique feature of this new technology is that the stabilization and reinforcement of the top-cloth are achieved without the use of fusible

Fig 40 : Polymer-paste printing for direct stabilization

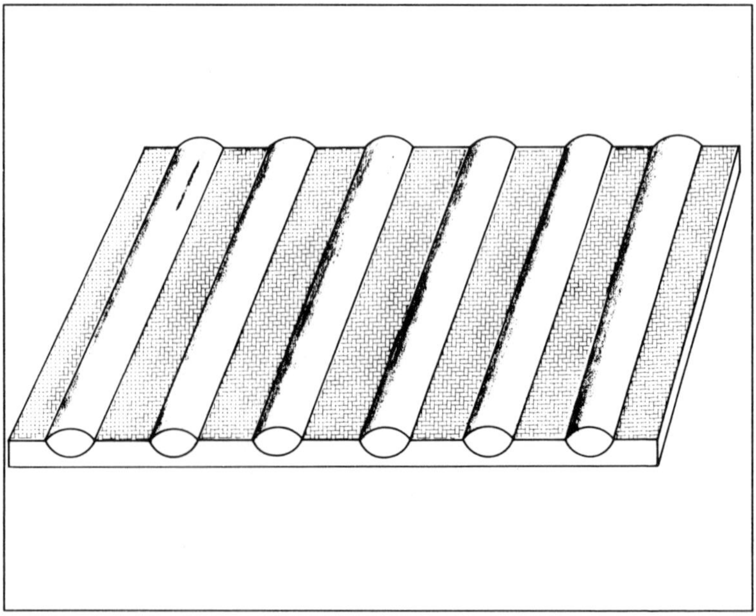

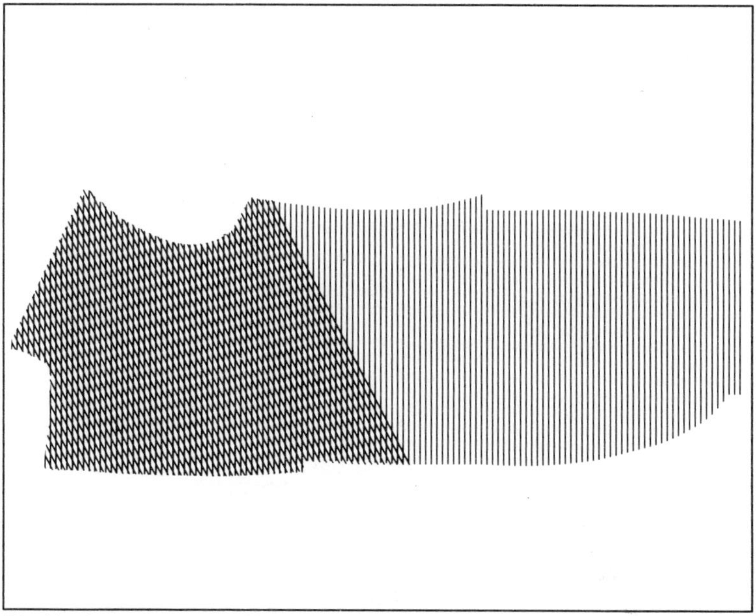

Fig 41 : Overprinting for additional reinforcements

interlinings. Instead, the stabilization material is printed directly onto the top-cloth. This process is known as direct stabilization, and its principles are as follows.

A polymer paste is printed onto the cloth in the form of closely spaced lines, and, when subjected to pressure, the lines penetrate into the cloth to about half of their thickness (*Fig 40*).

The paste is mixed with a polymer cross-linking agent to ensure an elastic construction of the printed line. The penetration and quantity of the paste are determined by the pressure of the printing unit, and this is automatically adjusted according to the cloth to be printed. There are a number of polymer pastes available, which enable a variety of handles to be produced, ranging from a full, strong handle to a smooth, soft handle.

The lines themselves can be printed in any direction in relation to the top-cloth, which thus provides more resiliency in specific areas of the garment. It is also possible to print over stabilized areas for reinforcement of a particular zone, such as the chest area of a man's jacket (*Fig 41*).

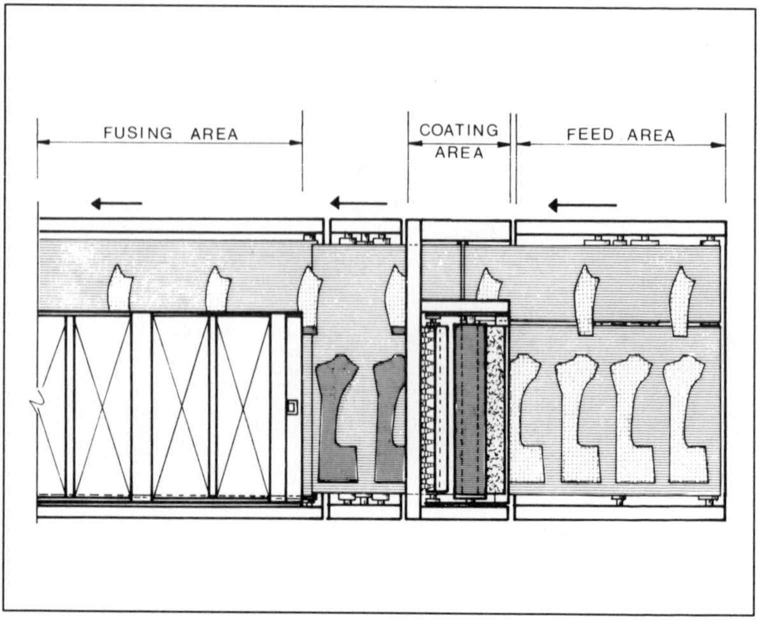

Fig 42 : Schematic work flow for direct stabilization

After printing, the parts are transported by conveyor belt to the drying and curing zone. This zone consists of three separate sections, each section being heated by infra-red radiators and convectors. The temperatures of the zones are controlled separately, which provides optimum conditions for drying and cross-linking. When the parts emerge from the last zone, they are transported to the cooling area and removed either manually or by a mechanical stacker (*Fig. 42*).

The configuration of the machine allows for a wide variety of printing applications, ranging from the entire area of a large part to a small strip across the cuff of a sleeve. In addition, the machine can be used for continuous fusing in case a specific type of applied reinforcement is required on top of the coated area.

This technology is relatively new but has a significant potential because of the advantages that it offers, such as:

(a) the savings in fusible interlinings.

(b) the elimination of nearly all the work connected with cutting fusibles;

(c) the reduction of storage space and logistic problems;

(d) versatility in the handling of different types of application.

These factors plus the resultant quality must be obviously weighed against the cost/performance of conventional fusing systems. If the current situation in the clothing industry is any guide, direct stabilization will play an increasingly important role in the future.

To sum up, the correct fusing of garment components is an essential prerequisite for garment quality. The machinery selected must therefore be able to perform this process in the most effective manner. This is one subject connected with clothing production in which the choice is not between cheap and expensive, but rather right for the job or not.

8. WORKING METHODS

The major component of fusing is heat, and all presses maintain a specific temperature irrespective of whether or not parts are in the heating zone of the machine. The continual consumption of expensive energy is thus a prime cost factor in the operation of a fusing press, and the objective of any working method employed must therefore be to utilize the energy consumption as efficiently as possible.

The operative elements of a working method for presses are:

(a) preparing the assemblies,

(b) loading the assemblies,

and then, after the fusing process,

(c) unloading and stacking the fused assemblies,

and in principle, these elements apply to both flat-bed and continuous fusing machines. The specific applications are as follows.

(1) *Flat-bed-tray Presses* With tray-feed machines, the most consistent fusing quality is obtained by cycling the press as often as the fusing-time cycle will permit. There is no production advantage in having the press closed for 15 s and then open for 3 or 4 min while the operator prepares assemblies directly on the tray. Consequently, maximum press utilization can be obtained only when a balance exists between:

(a) preparation of the assemblies on a lightweight card-board carrier;

(b) the cycle of

(i) loading the carrier onto the tray,

(ii) the movement of the tray into the fusing zone,

(iii) the complete fusing cycle, and

(iv) the return of the tray to its original position;

(c) stripping the carrier and stacking the fused components into their respective bundles; and

(d) returning one or more carriers to the preparation area.

The lightweight cardboard carrier is one of the most effective means of preparing assemblies for tray-feed machines. The carrier is the same shape as the fusing area, and the preparation operator first covers the carrier with the top-cloth components and then positions the fusible components on the cloth. Since the cardboard is semi-rigid, the carrier can be easily handled and transported without disturbing the prepared assemblies. In addition, the carrier also helps reduce the amount of resin build-up on the belts resulting from the effects of strike-back or strike-through. Some of the typical working configurations used for tray-feed presses are shown in *Figures 43, 44,* and *45*:

Fig 43: one single-tray press where the operator performs all the elements;

Fig 44: two single-tray presses where:

(a) one operator prepares the assemblies for both presses;

(b) one operator loads, operates, and unloads both presses;

(c) one operator loads, strips the carrier from both presses, and stacks the fused parts;

Fig 45: one twin-tray press with two operators, each of whom prepares, loads, and unloads alternate trays.

The preparation process is a highly labour-intensive operation, so the work stations must be engineered in order to provide the most efficient working situation.

(2) *Continuous Presses* The objective with this type of machine is to utilize the maximum area of the conveyor belt as it passes through the heating zone. This means that there should be the minimum space possible between the components when they are loaded onto the belt. Thus, the

Fig 43 : Single-tray press: layout

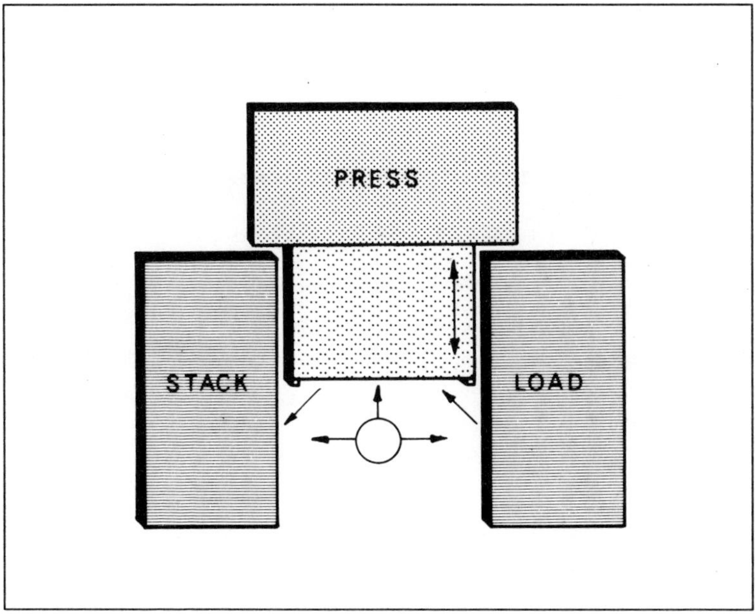

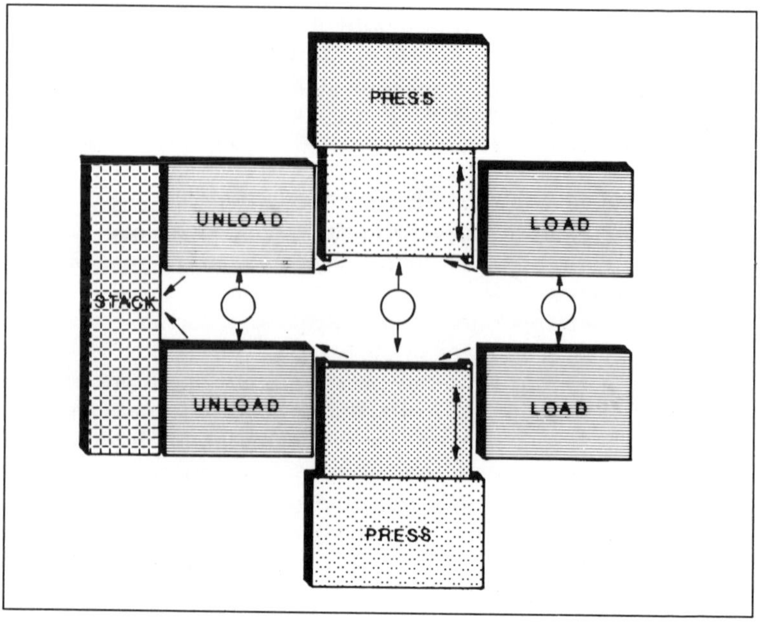

Fig 44 : Layout for two single-tray presses

Fig 45 : Twin-tray press: layout

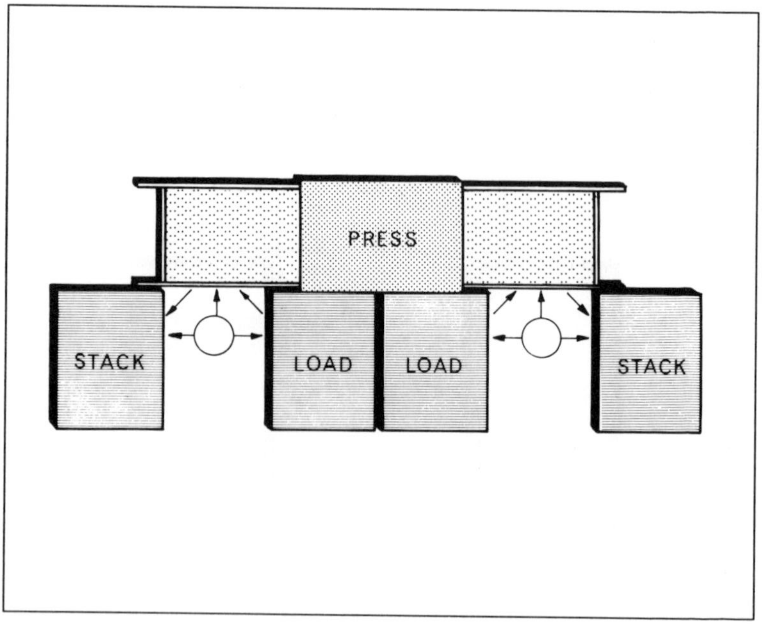

Fig 46 : Engineered work station

combined area-feeding rate of all of the operators must ensure the fullest utilization of area and consequently heat. Again, the importance of work-place engineering cannot be stressed too strongly (*Fig 46*).

A great advantage of continuous presses is that the unloading and stacking operations can be performed mechanically by a stacker positioned at the exit end of the press. The typical stacker has four modules, and the operating area of one module is equal to one-quarter of the conveyor-belt width. Thus, the belt is effectively divided into four parallel working lanes. Each lane, or combination of lanes, is allocated for the feeding of a specific component, where the deciding factor is the lane width required by the component. A four-lane feeding and stacking system can have the following combinations (*Fig 47*):

(*a*) lanes 1 - 2 - 3 - 4 individual lanes:

(*b*) lanes 1 + 2 combined; lanes 3 - 4 individual lanes;

(*c*) lanes 1 + 2 combined; lanes 3 + 4 combined;

(*d*) lanes 1 + 2 + 3 combined; lane 4 individual;

(*e*) lanes 1 + 2 + 3 + 4 combined.

Since the stacker is modular, it can be adjusted to function according to the lane allocation and length of the

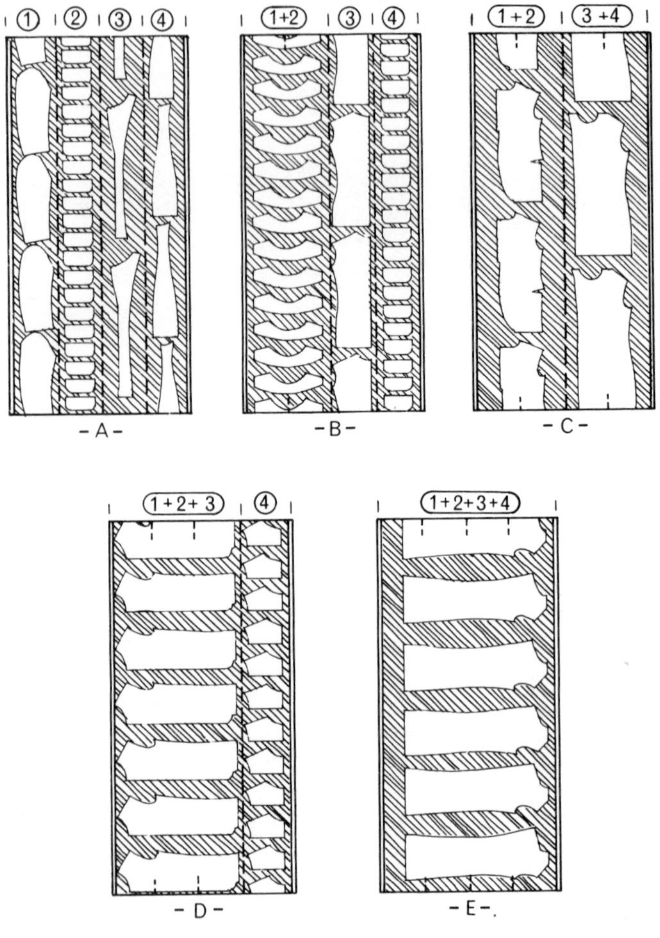

Fig 47 : Stacker lane combinations

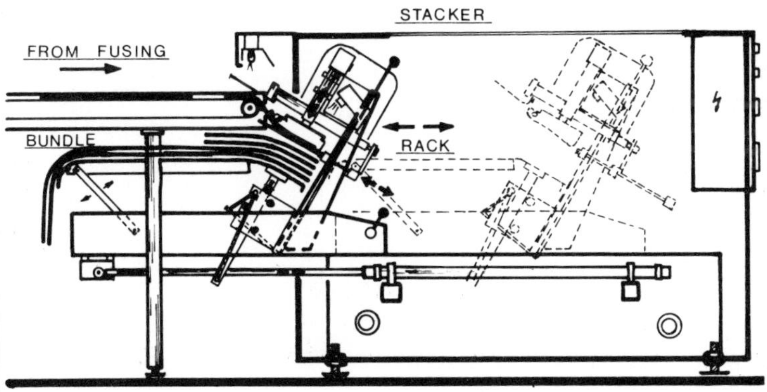

Fig 48 : Sectional view of modular stacker

component. The stacker itself works independently of the conveyor-belt speed and is actuated by photocells that 'sense' the position of the component. Thus the stacker removes and stacks components at the same rate as they are fed onto the conveyor belt. Work is removed from the belt by a simple grasping mechanism that positions the component onto a horizontal rack. This rack then swings backward, which causes the fused assembly to fall lightly onto the stacker tables. The rack returns to its starting position together with the grasping mechanism, ready for the next stacking cycle. When a particular bundle of components is completed, it is removed from the stacking table and transferred to a storage area. The basic elements of the modular stacker are illustrated in *Fig 48.*

A recent development for the loading area is a robotic system, which picks up and positions the fusible onto the top-cloth. The operator places the cloth component onto the conveyor, and the visual capability of the robot enables it to locate the correct position for the fusible while the cloth component is being transported past the work station. At the time of writing, the robotic systems available are capable of picking up and positioning only relatively small parts, such as flaps, collars, and cuffs, etc. However, it is reasonable to expect that, within a few years, the loading of cloth and fusibles could be performed exclusively by robots.

To sum up,one of the methods by which labour costs can be reduced is to increase the technological content of the production processes. This means that engineered work stations, automatic stackers, and eventually robots must become an integral part of the fusing process.

9. THE CONTROL OF QUALITY

All reputable manufacturers of fusibles maintain extensive laboratory facilities that continually test and develop their products. These laboratories also prepare data sheets that define the conditions under which a particular fusible will give its optimal performance. These data sheets are supplied to the clothing manufacturer and generally include the following information:

(a) the type or types of top-cloth that are most suitable to use with a particular fusible;

(b) the technical specification of the fusible, e.g., weight, composition, resin, and coating system;

(c) suitability for dry-cleaning or washing or both and the conditions to be employed;

(d) recommended applications of the fusible, e.g., components and function;

(e) residual shrinkage of the fusible;

(f) a matrix that gives the temperature, time, and pressure for recommended fusing equipment, such as:

(i) hand-iron,

(ii) steam press,

(iii) flat-bed press,

(iv) continuous press;

(g) the recommended grain direction of the fusible in relation to the direction of the top-cloth; and

(h) storage conditions.

Relatively speaking, fusible interlinings are precision products, and, as such, it is essential to ensure that they are fused on suitable equipment under strict control. Most fusible-equipment suppliers stress the importance of carrying out the following tests daily to ensure trouble-free fusing.

(A) *Temperature:* Two practical means of measuring temperature in enclosed areas are as follows.

(1) *Temperature Test Papers:* These are strips of paper that are calibrated in increments of 2°C and react to temperature by changing the colour of the segment with the appropriate temperature rating. Test papers are inexpensive and easy to use, especially for checking a number of points simultaneously. They usually have an accuracy of ±1%.

(2) *Pyrometers:* Although the accuracy of good-quality, rapid-response pyrometers is precise, they are costly, especially those designed for multi-point checking. In addi-

tion, they are difficult to use on continuous presses, since the thermocouple wire probe could possibly damage the conveyor belts.

It is generally accepted that temperature test papers are the cheapest and most practical means of measuring temperature under production conditions.

Temperature-testing Routine: All Presses

(a) The press should be switched on at least one hour before production starts.

(b) Checks should commence 40 min after the press has been switched on.

(c) The thermostats must be set to required levels.

Temperature-testing Routine: Flat-bed Press

The press should be set to operate 20 s closed and 20 s open, and this cycle should be repeated at least six times. This generates heat into the areas of the press that are not automatically heated, i.e., passing heat from the press heat to the buck. The temperature is checked by the following means.

(i) Two paper templates are cut to the size of the fusing area.

(ii) Temperature test papers are stuck to the templates in the twelve positions shown in *Fig 49*.

(iii) The template is placed in the press and heated for 20 s with the buck fully closed and then withdrawn.

(iv) The second template is inserted after an interval of 20 s and heated under the same conditions as the first test. The interval between tests is to simulate normal-production loading cycles.

(v) The results of both tests are compared to check if there are temperature variations at any one of the positions.

Temperature-testing Routine: Continuous Fusing Press

Temperature ratings are checked as follows.

(i) Ten sheets of paper about 20 cm long are cut to the width of the machine.

(ii) Temperature test papers are stuck to these sheets in the four positions shown in *Fig 50*.

(iii) The sheets are passed through the machine at 10-s intervals with a belt speed that would provide an equivalent fusing time of 20 s.

For All Presses

The maximum temperature variation between the points checked should not exceed 10°C, and each point should preferably be within ±5°C of the set temperature. Good temperature control is considered to be in the range of ± 2°C

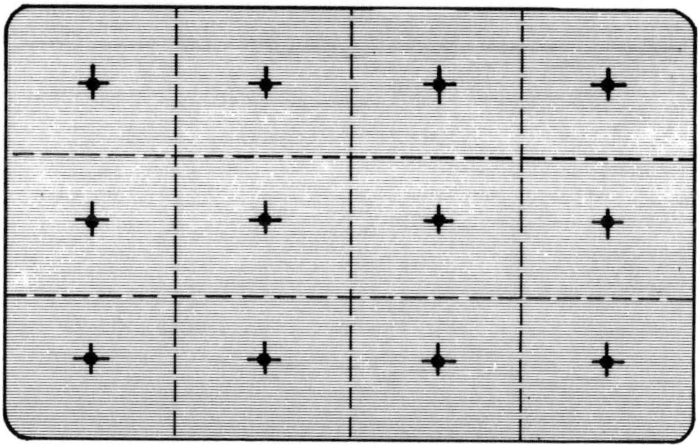

Fig 49 : Temperature-test positions: flat-bed press

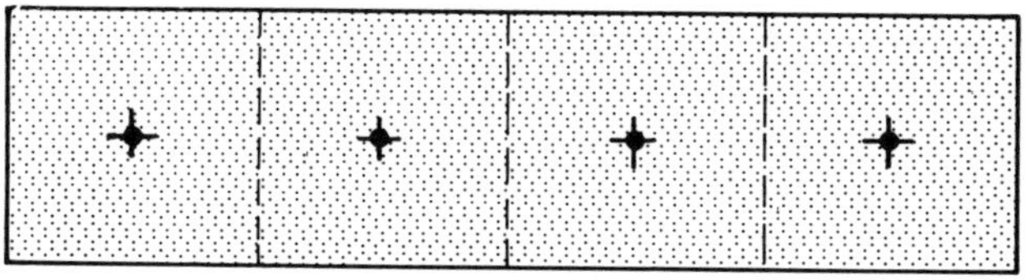

Fig 50 : Temperature-test position: continuous press

of the thermostat settings. If it is not possible to adjust the thermostats in order to obtain the correct range, then the press manufacturer or service agent should be consulted immediately.

All test readings should be entered in a log book. This will clearly highlight any trends in temperature variation that may not otherwise be apparent. The checking procedure must be carried out as follows.

(a) Immediately before production commences.

(b) After unscheduled breaks in production. With flat-bed presses, after breaks of as short as 5 min, the press

should be opened and closed six times to regenerate the lost heat before recommencing work. Continuous presses should be allowed to run for a least 10 min before starting.

(c) If the press has been inactive for more than 30 min (i.e., meal breaks, etc.), the press should be checked as at the start of production.

(d) If the press has been switched off for any reason during the working day, it should be allowed to warm-up and a complete set of readings taken.

(e) It is important that the press is sited correctly, since temperature can be significantly affected by sudden draughts. The location of fusing presses alongside open windows is not recommended.

(B) Pressure: Flat-bed Presses: The following procedure is used to check for the adequate and equal distribution of pressure over the buck.

(a) The pressure dial is set to the recommended level. Throughout the day, the operator should watch the pressure gauge to ensure that the specific pressure is consistently reached during the fusing cycle.

(b) There are two factors that can affect the equal distribution of pressure over the bucks: the flatness of the press head and the condition of the buck cladding.

(i) To check flatness, a metal straight edge is placed along the front edge of the press head. This process is repeated along the back edge and, finally, diagonally corner to corner. If any light shows under the straight edge, the press head is bowed. This is a major fault and should be referred to the press manufacturer.

(ii) To check that the alignment of the press head is correct, kraft-paper strips of about 15 cm x 3 cm are placed around the edge of the lower plate, about 4 cm of the paper being left protruding (*Fig 51*). When the press is closed, the strips should resist firm pulling. If the strips are easily removed, a fault exists in the pressure system.

(iii) An alternative test is to fuse a piece of cloth the size of the buck with an identically sized piece of fusible. After cooling, the assembly should be slowly stripped. In some cases, the fusible fibres will remain on the fused side of the cloth, and, at other times, the fibres of the cloth will be pulled out by the fusible. In either event, an obvious optical pattern will result, which

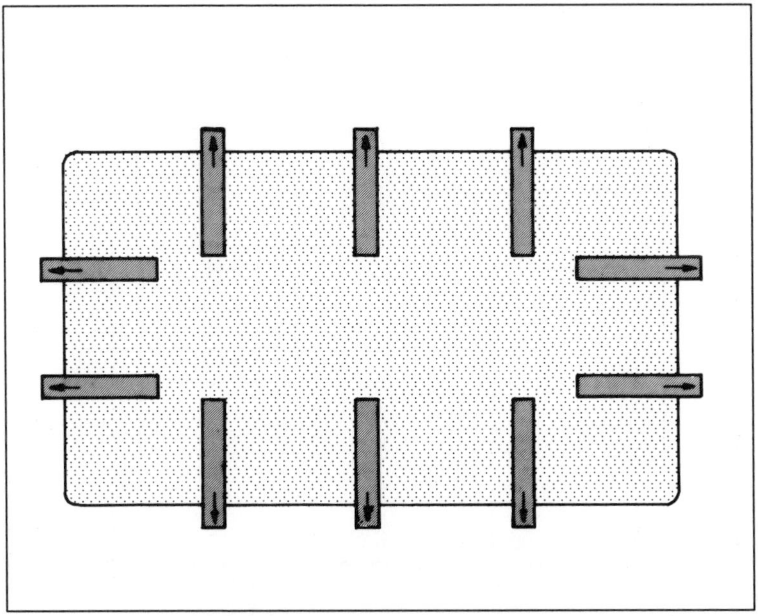

Fig 51 : Alignment testing with paper strips

clearly shows the pressure variations.

(iv) The condition of the buck cladding should be examined regularly and worn, damaged, or badly soiled materials should be replaced.

Continuous-fusing Processes: The pressure on this type of press usually relates to the air or oil pressure applied to the pressure system and not the actual pressure on the garment parts. The preferred test is as *(iii)* for flat-bed presses.

For all presses, the checking of pressure should be done at least once a week.

(C) Time

The 'time' given by the manufacturer's data sheet is the actual time period for which temperature/pressure is exerted. On a flat-bed press, pressure systems sometimes take from 3 to 5 s for the full pressure to be reached after the press is closed. On continuous presses, 'time' is measured from the entry of the assembly into the press until it emerges. Timers can go wrong, and the time cycle is best checked by the use of a stopwatch. Timing accuracy should be checked at least once a day.

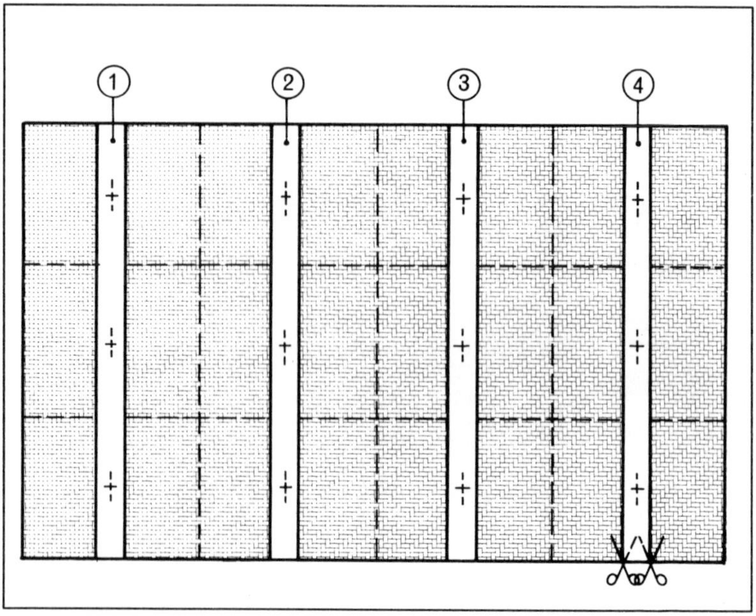

Fig 52 : Peel-strength testing

(D) Peel-strength (Adhesion) Test

High levels of adhesion are meaningless in themselves, and the garment manufacturer should know the minimum level of adhesion required to ensure that the bond between the cloth and fusible will satisfactorily withstand wear and the relevant methods of washing or dry-cleaning or both. Each fabric/ fusible combination will have its own minimum level, but it is usually possible to set one or two standards to cover the whole range of fabrics being fused in a particular factory. The most widely used adhesion test is the following. A piece of cloth, about the size of the press, should be fused with the relevant fusible. For continuous presses, its width should be equal to the width of the press and its length about 80 cm. Strips approximately 5 cm wide should be cut in the warp direction of the cloth, with each strip passing through the temperature test points (*Fig 52*). The strips should be subjected to 'pull' tests on a tensometer or a good spring balance, with the readings correlated with the temperature and pressure tests. Adhesion tests should be carried out twice daily.

(E) Dry-clean and Wash Test

It is recommended that garments be subjected to appropriate cleaning methods on a regular basis. One garment in each cloth/fusible combination should be tested either weekly or fortnightly.

(F) Records

A suitable scheme should cover the results of the tests for:
temperature,
pressure,
time,
adhesion levels, and
wash or dry-clean (or both) results.

In addition, a maintenance record should be kept for each machine. Operators should be provided with a simple data sheet listing fusing conditions for each top-cloth/fusible combination currently in production.

To sum up, the importance of regular checks on equipment and quality of fusing cannot be overemphasized. These checks will ensure trouble-free fusing irrespective of the type of top-cloth and fusible being used. Carried out properly, these checking procedures can prevent most major problems connected with the fusing process.

10. GLOSSARY OF TERMS*

The following terminology is commonly used in discussing the fusing process, fusing equipment, and fusible-interlining materials.

Base Material

Interfacing material on which thermoplastic adhesive is coated, printed, or sprayed. This material should have good shape-retention characteristics by itself, be washable and dry-cleanable, and not rely on adhesive to impart interfacing characteristics.

Bond Strength

The strength of bond between fusible and fabric. Many fusible suppliers rate this in terms of the force (in ounces force) necessary to separate a 1-in. strip of fused material.

Bubbling

An imperfect fuse resulting in the fusible not adhering to the garment section over the entire area of the fabric, usually caused by uneven temperatures or pressures, or by under- or over-fusing.

Colour Change

The temporary or permanent change in colour of a fabric caused by the action of heat on certain dyes.

Compatible Shrinkage

Shrinkage test of fused interfacing to outer fabric to ascertain if a fusible has the proper amount of shrinkage. Shrinkage should not cause delamination or bubbling of the combined package. (This test must be performed after fusing only, and the test should not be made separately.)

Cooling Time

The time necessary for the resin to cool and set.

Delamination

Complete breakdown of bond between fusible and fabric.

Dwell Time

See Fusing Time.

Fully Fused Front

Use of a fusible interfacing over the whole forepart of a man's or woman's coat.

* *With acknowledgements to Union Special Corporation, Chicago, U.S.A.*

Fusing Machine
Equipment designed specifically for fusing, having rigidly controlled temperature and pressure tolerances, readable pressure gauge, and adjustable temperatures, pressures, and time cycles.

Fusible Chest-piece
A fusible interfacing with softness, resilience, and bulk, used to give shape in the chest area of a garment.

Fusible Interfacing or Interlining
Woven or nonwoven base material, printed, sprayed, or sintered with fusible adhesive.

Fusing
The bonding of interlining to a garment section by the application of heat and pressure.

Fusing-cycle Time
The complete time for the fusing operation, including loading time, fusing time, and unloading time.

Fusing Pressure
Pressure (in lbf/in^2 or atm) exerted by the head or rollers during a fusing cycle. It is equal to the total force exerted divided by the fusing area and is recommended by the fusible supplier as being appropriate to properly fused interlinings for a selected fabric.

Fusing Time
The time for which the head remains closed, controlled by a timer usually stipulated by a fusible supplier as being requisite for the proper fusing of the interfacing to a selected fabric. For a continuous conveyorized fusing machine, this is the length of time it takes the garment piece to pass through the machine.

Hand or Handle
A term used to describe a number of factors, including the degree of shape, stiffness, or softness obtained after fusing to the topcloth.

Handling System
Apparatus associated with a fusing machine to transport the garment section in and out of the fusing area in an automated or semi-automated fashion. Examples of handling systems are: conveyorized fusing machines and double- and single-tray fusing machines.

Loading
The act of placing a garment section on a loading area of a handling system and then placing the proper fusible on top of the garment section.

Over-fusing
Excessive heat, time, or pressure applied to the fusible so that the resin migrates completely into the garment fabric and leaves a weak bond between the fusible base and the top cloth.

Paper Tray
A sheet of lightweight yet rigid paper (e.g., stiff pattern paper), cut to the size of the fusing area, on which garment parts and fusibles are pre-loaded for insertion into the fusing machine. Particularly advantageous for small parts.

Pressure Indicator
A gauge device that indicates fusing pressures.

Pyrometer
A portable measuring device for sensing the temperature at various points on the fusing-machine head.

Resin Transference
The spoiling of a garment or garment part by resin particles from a dirty machine head.

Sandwich Fusing
The laying-up of garment parts to be fused one on top of the other. The fusible interlining for each piece is sandwiched between the garment pieces in such a way that both garment pieces are fused simultaneously and independently.

Stay Tape
A bias- or straight-cut strip of fusible used to give control and stability in particular parts of garments.

Strike-back
The appearance of adhesive material on the non-adhesive side of the fusible after fusing. Strike-back causes sticking of sections and accumulation of adhesives on the head of the fusing machine.

Strike-through
Appearance of adhesive on the outer face of the fabric being fused. It is usually caused by too long a fusing time or to high a fusing temperature or pressure.

Temperature Differential
Difference between the highest and lowest temperatures on a head for a given temperature setting measured over surface and over time.

Thermopaper or Pyrometer Paper
Chemically treated paper strips that react to temperatures by changing colour at a predetermined rating. These papers are available in various temperature ratings and are inserted into the fusing machine under actual production conditions as a means of checking the actual fusing temperature. Recommended for use on conveyorized fusing machines.

Under-fusing
Insufficient heat, time, or pressure applied to the fusible interfacing, so that the adhesive is not activated and does not adhere to the garment fabric.

11. BIBLIOGRAPHY

C.M.C. Dorkin and N.H. Chamberlain. 'The Practical Performance of Fusing Presses' (Clothing Institute Technological Report No. 19), Clothing Institute, London, 1969.
C.M.C. Dorkin, N.H. Chamberlain, R.M. Crow, and A. Kaltsoyannis. 'The Testing and Utilisation of Fusible Interlinings', Clothing Institute, London, 1968.
J. Scheines. 'Apparel Engineering Handbook', Kogos International, U.S.A.
'Technology and the Garment Industry', National Economic Development Office, London, 1971.
H.C. Carr. 'The Clothing Factory', The Clothing Institute, London, 1972.
A.J. Hall. 'Standard Handbook of Textiles', Heywood Books, London, 6th edition, 1965.
H. Rose. 'Fusible Interlinings – Origins and Technology', Highgate Publications, London, 1979.

Journals and Papers
Technical Bulletins published by the British Interlining Manufacturers' Association (B.I.M.A.)
Clothing Institute Journal (England).
Bobbin (U.S.A.).
Manufacturing Clothier (England).
R.A. Burfitt. *Technical Requirements for Fusing Techniques.*

ACKNOWLEDGEMENTS

My grateful thanks are also due to the machine producers, fusible manufacturers, and agents who kindly supplied me with all of the materials and answers that I requested.

Machine Producers

Kannegiesser - Vlotho, Germany (BRD).
Reliant Tooling Company Ltd, Chesham, England.
Hashima Ltd, Gifu-City, Japan.
Vaporetta GmbH, Ascheffenburg, Germany (BRD).
Paul Kiefel GmbH, Freilassing, Germany (BRD).
Meier GmbH, Munich, Germany (BRD).
Union Special Corporation, Chicago, U.S.A.
Emig KG, Krefeld, Germany (BRD).
Gematic, Villeurbanne, France.

Manufacturers of Fusibles

Staflex International, England.
Carl Freudenberg, Germany (BRD).
Lainière de Picardie (UK) Ltd, England.
Crown Textile Corporation, U.S.A.
DJH Industries, U.S.A.
Kufner Textile Works, Germany (BRD).
Haensel Textile Company, Germany (BRD).
Delius & Sohne, Germany (BRD).
Helsa Werke, Germany (BRD).
C.F. Plouquet, Germany (BRD).

Agents

A. Silberstein Ltd, Tel-Aviv, Israel.
Dov Shafir Ltd, Ramat-Gan, Israel.
Schleien Bros Ltd, Tel-Aviv, Israel.
L. Taube Ltd, Tel-Aviv, Israel.

Fig 7 : Fusible application : rainwear

Fig 24 : Pressure system of continuous press

Fig 25 : Types of flat-bed press

Fig 38 : High-frequency fusing press — work flow

Fig 51 : Alignment testing with paper strips